The Virtual Pagan

This book belongs to...
Lee Joanne Harper

If this book should care
to roam. Please smack
its bum, then, send
it home!

The
Virtual
Pagan

Exploring Wicca and Paganism
through the Internet

Lisa McSherry

WEISERBOOKS
Boston, MA/York Beach, ME

First published in 2002 by
Red Wheel/Weiser, LLC
York Beach, ME
With offices at:
368 Congress Street
Boston, MA 02210
www.redwheelweiser.com

Library of Congress Cataloging-in-Publication Data

McSherry, Lisa.
Virtual Pagan : exploring Wicca and Paganism through the Internet / Lisa McSherry.
p. cm.
Includes bibliographical references.
ISBN 1-57863-253-6 (pbk.)
1. Occultism—Computer network resources. I. Title.

BF1411 .M185 2002
025.06'13343—dc21
2001046897

Typeset in 9.5/13 Celeste

Printed in Canada
TCP

09 08 07 06 05 04 03 02
9 8 7 6 5 4 3 2 1

The paper used in this publication meets the minimum requirements of the American
National Standard for Information Sciences-Permanence of Paper for Printed Library
Materials Z39.48-1992 (R1997).

Contents

/part_one/
Beginnings, 1

/part_two/
Interactions, 35

/part_three/
Practice, 85

Appendices, 129

Acknowledgments

The creation of a book is primarily a solitary endeavor, although many people supported me in my creative journey.

First and foremost, Lady Mystara—without your dream, this book would not be. To the members of ShadowMoon and JaguarMoon covens—without the generosity and magick of my coven sibs, we would not have been the success that we are, and I would have nothing to say. In particular, River Rock assisted in clarifying my thinking about anonymity in cyber circles.

Corynne McSherry, Kathleen McGill, and Karla Ziegenbalg provided me with ongoing analysis, critique, support, and guidance. Ladies, from the bones of my fingers, I thank you.

William Grimaldi provided humor therapy, encouragement, and introduced me to Anthony Barrata, who came along at a fortuitous moment and solved my biggest technical problem.

I am also truly grateful for Jan Johnson of Red Wheel/Weiser Books. She took a chance on an unknown author, supported my slightly skewed vision for this book, and worked with me to produce a far better book than I would have otherwise.

In the beginning and at the end, there is always Michael Compton. I may be better at words, but you have always been better at gestures. Kisses and the Moon.

Introduction

The Virtual Pagan is a practical guide, based on my experiences in pagan groups and online, over the last two decades of being a witch. Drawing on my knowledge, as well as that of other pagans on the Internet, this book is a reference and guide to the many aspects of getting online, choosing, and then participating in an online coven or other magickal group.

The Internet community is constantly evolving, as is the material in this book. I have written this book to act as a counterpart to my Web site, www.thevirtualpagan.org. This Web site is a clearinghouse for further information, updates, and modifications to the knowledge that you find in these pages. The book also contains references to various places on the Web where you can find additional sources of information about a particular topic.

The Virtual Pagan is divided into three parts:

▸ Part One, "Beginnings," focuses on getting connected to the Internet and starting your exploration;

▸ Part Two, "Interactions," explores meeting other pagans online and learning how to communicate with them in a healthy manner;

▸ In Part Three, "Practice," we will begin to prepare for and enact rituals online.

The Appendices contain useful information on how to organize your information, a glossary, resources, and more.

The first chapter orients you to Cyberspace, magick, and some of the definitions that you need to know. For example, I discuss covens throughout this book. The term includes any group of people who assemble for a magickal purpose. What I call a coven, others sometimes call a grove, or temple, or other similar terms.

Chapters 2 and 3 guide you through the technological aspects of a cybercoven: minimum hardware requirements and software recommendations so you can gain online access. Technology is the area that changes most quickly. Coming out of the broom closet is an intensely personal decision and when you interact with other people online, the manifestation of your decision might become an issue. Online privacy and anonymity can be maintained, and Chapter 3 explains how.

"What if I am only interested in joining other pagans online?" you might ask. Excellent question! Chapter 4 should help you examine magickal groups in the physical and cyber realms, and decide whether these groups are right for you. The occult community is replete with frauds; hopefully this chapter will help you to spot them. Meeting with a group for the first time can be scary, but my explanation of circle etiquette on page 48 should ease your trepidation.

Chapters 5 and 6 discuss the ongoing interactions of successful groups online. One of the most important issues for a cybercoven is to communicate well in a text-based environment. Negative interactions are what shut down most physical covens, and the potential for miscommunication is multiplied online. The effect of words in print and the speed with which we can answer perceived insults can create intense emotions and conflict. Knowing when to step in and prevent an imminent flame war is a useful skill; not getting involved in one is an even better strategy. An old Tai Chi master once observed that, when you offer resistance or opposition, you are the one who creates the conflict.

Chapter 7 takes an honest look at the problems of maintaining a well-functioning group online. Politics are one of the dirty secrets of the Craft, and Cyberspace has more than its fair share of strong personalities. I will discuss the downside of gathering with other pagans, from the disappearing members to conflict resolution, and how best to cope with problems as they arise.

Chapter 8 explores the new concept of online ritual. In this chapter, you will learn the basic building blocks of magick: meditation, visualization, and physical preparation. No matter where you perform your ritual, with a group or as a solitary, online or physical, the ritual's success depends on your mental state, also called your intent.

Chapter 9 contains a variety of rituals with examples of the differences between physical and cyber rituals. Any traditional ritual can be performed online, including initiations. The examples encourage you to explore further.

/Your Author:

As a practicing witch since the age of fourteen, when I discovered my mother's copy of *The Spiral Dance*, I have had a variety of ritual and magickal experiences. Prior to being a member of ShadowMoon, I practiced as a solitary for seven years, although I occasionally attended rituals hosted by the magickal group New Moon New York and took classes through the Open Center in New York City, where I was then living. Before New York, I participated in a variety of circles and magickal classes in and around Davis and Sacramento, California.

In 1997, I joined ShadowMoon, of one of the oldest cybercovens in existence. With twenty to sixty members from all over the globe, we were pioneers in exploring and understanding the practice of Paganism in Cyberspace. Our primary mission is to teach Wicca, and more than forty members have graduated over the years. Some coven members have joined us permanently; others have gone on to found their own covens and teach. With the blessing of our High Priestess, I left in 2000 and formed my own cybercoven, JaguarMoon.

I have worked in eclectic teaching circles and hierarchical covens. I taught basic magick to people I met through classes, as well as to like-minded friends. I have also had many teachers, each of whom provided parts of the wisdom in this book.

I do not speak for all witches, nor do I speak for all Wiccans. I do not even speak for the coven that I lead. I speak for myself, and this book is a reflection of my views, attitudes, and beliefs. Take what I say, write, and show you, and think about this knowledge. Contrast it with your own beliefs and practices, how you live, and your own values. Use what works for you; adapt the things that you can and ignore the rest. You have the mind and skills the Lord and Lady gave you. Use them.

Now, come, I invite you to join me in this exciting new adventure. Let's get started with making you a virtual pagan!

/part_one/

Beginnings

1: Orientation

Humming under your breath, you thoroughly clean your sacred space, preparing it for the ritual. Your ritual tools are in their appropriate places. The incense is ready to be burned, the chalice filled with fresh water, and you are in the right frame of mind.

You check the time, and then shower. A bunch of rosemary tied to the showerhead fills the air with its sharp scent as the day's residue washes away, down the drain. You are cleansed, in mind and body.

Taking a deep breath, you reenter the sacred space. Holding your athame, facing east, you call out, "All Hail the Guardians of the East!" Facing each direction in turn, you call the Quarters, creating the magick circle, the sacred space between the worlds.

Sitting at your altar, you take a deep relaxing breath and begin to type. The ritual may now begin.

<Welcome to sandnet.org

`</join #thetemple`

<Maat has joined #thetemple

Maat> `Greetings everyone!`

Brightayes> `Hi Ma'at!`

ThunderBunnie> `Hi Ma'at!`

Lady_Jaiyne > Hi Ma'at!

BullSpawn> Good evening Ma'at!

GorgonsDaughter> Hi Ma'at!

Imago> Hey Ma'at!

Brightayes> Hugs Ma'at. Good to see you.

Maat> Looks like we're all here——shall we
  begin?

Maat changes her nickname to Lady_Maat.

Lady_Ma'at walks to the East and raises her
  athame high.

Lady_Maat> Mighty Mother! Strike this blade
  with light

Lady_Maat> that I may cast the sacred circle
  between the worlds!

And so the ritual continues.

Welcome to the next step in pagan evolution—the cybercoven. Where once we were prevented from reaching out by fear of persecution, forced to practice as solitaries, we are now free to worship in Cyberspace. No longer limited by geography in our search for kindred pagans, we are limited only by the speed of our computers and our ability to use them. We use new tools: the computer, the modem, the Internet, and the Web, along with our athame and chalice. We are cyberwitches, riding the energies of a new dimension, Cyberspace.

WHAT IS CYBERSPACE?

As Jennifer Cobb writes in *CyberGrace, the Search for God in the Digital World*, "The reality of Cyberspace transcends the dualism represented by objectified mind and matter. Cyberspace is a messy and complex world of experience, both objective and subjective.... It has the potential for opening us to a new way of experiencing the world, a way that relies on a divine reality to give it meaning and substance."[1]

Going online, we immerse ourselves in a nonlinear environment, one that places us in a reality where we control our movements, while being transported to places unseen and unimagined. There is no tidy, rational way to move through Cyberspace. It is an environment of loops and links, where everything is connected in a seemingly infinite network or web.

The nonlinear environment in Cyberspace is like the real world if we visualize the physical world as a vast, complex flow of energy and movement. Within this flow, everything—from an atom to Mt. Rainier—participates in its own process. Everything is interwoven in a vast pool of being and becoming. Pagans have seen the world in this way for centuries. The intimate knowledge of this energetic connection of all with all is what allows us to do magick, to recreate our physical reality in accordance with the powerful energy of Divine will.

Now we have created that world within and through our computers. We created it, we shape its evolution, and we manipulate it. Yet, an element remains beyond our control, a creative energy that accepts our designs and then transforms them into something we did not expect. As Erik Davis, author of *Techgnosis: Myth, Magic, and Mysticism in the Age of Information*, says, "The computer is the most animated and intelligent of machines, the most interactive, and by far the least 'mechanical.'"[2] Cyberspace is a technological doorway to the astral plane. The entrance cannot be found in a piece of a computer or in a program that sits unused on your desk. Cyberspace is what happens when you join software and hardware and then activate them. By the time our conscious minds view the products created in the cyberplane, the process itself is already complete. Once we enter Cyberspace, we are no longer in the physical plane; we literally stand in a place between the worlds, one with heightened potential to be as sacred as any circle cast on the ground.

Cyberspace is a different dimension of interaction: a window between you and the other person who is typing. If you are familiar enough with your computer, you can project yourself into that realm. I associate myself with the words I type, seeing them as objects, rather than as symbols in my head. It is not just putting letters onto a screen; but describing an experience I am having. When we prepare for ritual, we alter our state of consciousness to one in which we can more easily access the astral plane, the place where energy is more accessible to human beings. Whether we alter our consciousness through solitary meditation, while holding the hand of a coven member, or while typing on our keyboards, we access the astral realm. As my partner said to me, "Cyberspace is just a techno-term for the astral plane or any other nonphysical reality. It is where you work [magick]."³ Cyberspace is our common ground, our energetic link to one another, like those achieved by chanting or drumming.

Because we understand that the extraordinary power of ritual arises from the Divine presence in all human beings, pagans embrace a philosophy of transformation—of the self and reality. We believe that it is our sacred task to manifest the Divine power we each possess for the greatest good within the world. Whether we call it Karma, or the Threefold Law, we also believe that what we do in the world will return to us, and so we strive to behave in an ethical manner at all times.

Whether our tools are the athame, a circle of sacred space, or a computer to access the Web, pagans love technology. And I argue that all tools are technology. The tinkering, the arcana of our religion, is its strength; we see nothing as set in stone, immutable, or unchanging.

For all that our roots are firmly planted within the rich loam of the past, pagans are well represented in the fields of technology. It may be that, as Arthur C. Clarke said, "Any sufficiently advanced technology is indistinguishable from magic."⁴ We use our tools of technology and often have no idea how they work; for example, how is it that your computer understands that a particular keystroke is equivalent to a specific symbol? This lack of direct knowledge makes the process of writing a letter on your computer a magickal experience. The logic of technology has become invisible—literally, occult. Then again, these glorious technologies are also magickal because, with their aid, we now can impress our will upon the stuff of the world, reshaping it, at least in part, according to our imagination.

DEFINING MAGICK

The ability to think seems to set us apart from animals. And, although we are concerned with living in the physical world, we are mental beings, thinking all the time. We plan, we brood, we get depressed or elated; all of it is a mental process. But the universe is mental too and, if we control our thinking, we see magnificent results in the everyday world.

Over the ages, many systems have developed to help us control our thoughts in an attempt to make us into better people. The practice of magick is one of the oldest of these systems. Magick is the study and application of psychic forces, using mental training, concentration, and a system of symbols to program the mind. (I spell "magick" with a *k* to distinguish it from the magic of today's modern illusionists with their entertaining sleight-of-hand tricks.)

The universe is pure energy. All that is seen and unseen is energy; science has taught us this. But the Vedic sages of India, the European Druids, and, after them, the witches, recognized this fact as well. Even the most solid stone contains millions and billions of atoms and molecules that orbit each other in an endless and graceful dance of Lifeforce. The electromagnetic energy that holds electrons, protons, and other microscopic particles in place has had many names: *Chi* (Asia), *Mana* (Polynesia), *Orenda* (Iroquois), the Force (Star Wars), and psychic energy (contemporary). I call it simply *energy*.

Energy is not static or inanimate; it is responsive and dynamic. It is fluid in its movements, yet it is also similar to light in its effects. The shape that energy assumes creates the pattern of the physical world that we see; all physical forms are energetic structures. We interact with this energy every day, in every moment of our lives. It constantly transforms, renews, or changes its shape within and around us. This constant change responds to and is driven by our thoughts and emotions in ways of which most of us are unaware and few understand. Energetic phenomena can manifest in reaction to our unconscious beliefs and emotions, as automatic and out of our control as our unconscious can be.

But when we focus our conscious mind on this process, everything shifts. Rather than a process out of our own conscious control, the shaping of energy and thus of the world becomes a precise and deliberate skill that is in our hands.

Magick is the art of consciously directing energy through focused will and effort to affect the things around us. Magickal practitioners effect these changes from the deeper levels of their consciousness. Not everyone can do it, and it is not an easy technique to teach. Magick is done from a higher level of consciousness, what some call the Higher Self.

Scientists who study people who have psychic abilities have found that when a person enters a psychic trance, his or her brainwaves change. Psychics who are in a trance state do not use the normal *beta* waves associated with ordinary, conscious thoughts; they use the *theta* and *delta* waves that are associated with sleep. This difference in mental state also holds true for a person who performs an act of magick; a change in consciousness occurs that affects the brainwaves. It is not difficult for a devoted practitioner to reach this level; but it is difficult to learn to do it at will, to be able to access this ability when needed. An accomplished witch can do it instantaneously. The student of magick, however, will take time to change his or her consciousness, and may have to work hard to master it.

Practitioners can perform magick in many different ways. All of them have the same basic goal: to focus energy from a state of higher consciousness and direct it. Visualization, trance, spellcraft, rituals of various sorts—all of these and many other techniques can create the necessary shift in consciousness.

Thus, the human mind and body appear to broadcast psychic energy or force, much like a radio station. (Kirlian photography and other parapsychological tests tend to support this theory.) This psychic force is the energy behind psychic phenomena and magick. Such a force does not exist in the form of a radio wave, since it behaves somewhat differently. The psychic force seems to be too subtle to be easily measured, but it also seems that everyone has some psychic ability.

Our consciousness is a spectrum and what we think is our core personality is actually only a shadow of our true self. Entering the silence and listening is the essence of magick. Theatrics and shenanigans are simply window dressing. Ultimately it helps us to do what we fight hardest not to do: wake up, see the truth, and be ourselves.

Magick is building community, teaching others, expressing the art that comes through your individual perspective, speaking truth, and sharing human existence.

Magick, witchcraft, alchemy, or any occult field, are complex subjects. Magick includes them all; it is a philosophy that has, as the late Aleister Crowley wrote, "The method of science—the aim of religion."[5]

BEING A PAGAN OR A WITCH

Many sources of information about Wicca and Paganism abound—much of it valuable, some of it contradictory. "Witch" comes from the Anglo-Saxon *wicce*, which derives from an Indo-European root word meaning "to bend or change" or "to do magic or religion." Therefore, the word is related to "wicker," "wiggle," and even "vicar." Related words are "pagan," which simply means "a country dweller," and "heathen," which means "a dweller on the heath." Both words came to mean "not Christian." Pagans were the European equivalent of Native Americans and other indigenous, nature-worshiping peoples. And, as further clarification, "warlock" is not the term for male witches; rather, it comes from the Germanic word for oathbreaker. The terms "witch," "Wiccan," and "pagan" are interchangeable throughout this book.

A witch is someone who focuses his or her will through a change in consciousness. Witches, however, do not affect physical reality solely through mental power. For instance, I can spend all day visualizing my athame rising off my altar and floating over to me; but, if I want to cut a circle, it is a lot easier to just pick up the tool. When witches use magick, we are not trying to influence matter; instead, we influence events. Events are matter in motion and matter in motion, as the practitioner of martial arts knows, can be deflected by a slight application of force. Complex events, therefore, can be profoundly altered by minor, intangible factors. This arcane knowledge has become the recently discovered field of chaos theory.

Wicca itself is a religion of initiation. Wiccans do not convert people to their point of view or way of life; rather, we indicate a direction that has been helpful to others in the past. Wicca is a progressive religion, with each new initiate adding to the pool of knowledge, expanding the pathways of the soul and, most importantly, intimately understanding himself or herself. Essentially, Wicca is spiritual training that develops the self to its highest and best potential.

If you are new to Paganism, then you need to know a few things about this religion:

1. We all truly only agree on one thing: "An' it harm none, do what ye will." As a result, we do not take any action—magickal or otherwise—that would harm any person, including ourselves.

2. We do not have a central authority, either document or person, to whom we all look. Instead, we each seek our own understanding of the world and our own path to follow in the world. Every practitioner of Wicca is a priest or priestess; we need no intermediaries between the Divine and us.

3. Wicca is not a religion of the masses. Initiates undergo constant self-examination and growth, striving for years to master a concept or technique. We do not turn away from the negative aspects of ourselves, but instead seek to understand and improve them.

4. Pagans revere the Earth as the Great Mother, Gaia, who sustains us all. All things in Her are sacred, including all of us. In this reverence, we mirror seasonal cycles within our own lives, greeting the stark beauty of winter with the same joy as the feasts of summer.

In Paganism, the term "coven" includes groups that also call themselves circles, groves, or temples. A cybercoven is a group that has most, if not all, of its interaction online. Although some socializing might occur in the physical realm, rituals, teaching, and conversations are held online. In contrast, a physical coven may use the Internet or the World Wide Web (WWW for short) to facilitate coven interactions, but most of its activities and rituals occur in the physical realm. (I will explain more about how the Internet and Web interact with one another in Chapter 2).

Some readers may ask, "Why join a coven?" An obvious answer is that there is power in numbers and, hence, a coven raises more energy. During rituals, the collective efforts of a coven are greater than those of a solitary practitioner. Group membership also connects you to a spiritual community of people who have similar theology and can provide general love and comfort. Joining a magickal group exposes you to a wider variety of beliefs and more information than you could probably find on your own. It is also far more difficult to misjudge your own progress when you have others to encourage you and measure yourself against. A coven can also provide support as you face the challenges that life hands you. Covens offer members challenge, intellectual stimulation, and magickal explorations.

Although a coven is generally described as a group of three or more witches who meet in an organized fashion, I also use the term to denote any group that meets for a magickal purpose. A coven is more than just a group, however. It is an intentional family, a community of people whom you choose to have in your life, from whom you learn, and with whom you can explore the inner and outer worlds. Such exploration does not happen in just a month or two. Perhaps blood is thicker than water, but love is stronger than the first two put together. Covens evolve over time. Your coven is with you for the good and the bad.

Being a cybercoven member is not for everyone. If you have a hard time learning without some kind of face-to-face communication, you might feel lonely and out of touch in a cybercoven. If you are really bad at answering e-mails, this may not be the place for you. However, if you like to read or meet new people with different points of view and experience, then membership in a cybercoven can be an exciting experience.

2: Starting Your Journey Into Cyberspace

Many people who view themselves as intuitive, psychic, and creative, operate from the right hemisphere of their brains. To them, the realm of computers reeks of the math and science that the left hemisphere of the brain processes. The left hemisphere is uncharted waters for the right-hemisphere crowd. But, while computers can be intimidating at first, Cyberspace is wondrous and offers a new arena for spiritual growth and exploration.

Look at some of the good things about Cyberspace:

▸ Would you like to share your ideas with other pagans? Join a general pagan newsgroup or mailing list;

▸ Are you looking for recommended reading lists? Use a search engine on the World Wide Web;

▸ If you need information about a specific Wiccan tradition, there's probably a Web site, newsgroup, or mailing list devoted to that subject;

▸ If you cannot find a local place to buy altar supplies, books, or other pagan essentials, you can shop online through a variety of Web sites, auctions, or catalogs;

▸ The Internet has an answer for every question. You can even use e-mail (electronic messages) to correspond quickly with friends, compared to snail-mail (the U.S. Postal Service), at a fraction the cost of telephone calls;

▸ Even if you live in the middle of the Bible Belt, you can enjoy daily contact with a thriving online pagan community.

This chapter explains the technological tools required to link to and explore the Internet and WWW. If you are comfortable with computers and have already begun to access the Internet, then just skim this chapter; or skip it entirely. However, you may find useful advice here about using computers as communication tools.

As the Information Age moves along at ever-increasing speeds, making the newest ideas obsolete almost immediately, some of the technological information in this chapter may be outdated by the time you read it. If so, just use it as a starting point. Because I am most familiar with PC technology, that is the terminology I use. However, if you and your group prefer Apple products, or Linux, or UNIX, please use them instead. The information in this chapter is not an endorsement of certain products over others. Use what you like, what you know, and what you think works best.

HARDWARE BASICS—THE EASY PART

Few rules exist for having the best piece of computer hardware. But buying a computer with a lot of memory, as much storage as you can afford, with the fastest processor and modem is a good start. The most basic implement is a personal computer with a moderately fast processor. Although some cybercoven members have older computers, a fast processor facilitates coven communication. The more random access memory (RAM) that you can afford, the better.

Here's a list of information about computer equipment with which you can get started:

▸ The processor is the part of the computer that governs how quickly it is able to handle commands;

▸ RAM is memory. It contributes to the speed of opening and closing programs. The more RAM you have, the less time your computer spends rearranging its memory for optimal performance.

Besides a processor, a monitor is needed, as is a good-sized hard drive, but perhaps the most important tool is a fast modem connection—at least 28.8

kilobytes per second (KBps). Here is a list of "rock bottom" basics:

▸ 14″ color monitor

▸ 16 MB (megabytes) RAM

▸ 1 GB hard drive

▸ 166MHz processor

▸ 1 MB video card

▸ 28.8 KBps modem

Components such as these can get you connected into the Internet. Great computer systems are available for somewhere around $500 (or less), and it is possible to put together a much nicer system than the basic one listed. My personal computer wish list would include:

▸ A faster modem (56 KBps) or a cable/Digital Subscriber Line (DSL) connection to maximize online speed

▸ At least 256 MB of RAM

▸ A huge (20 GB or more) hard drive

▸ A 1 GHz processor, or faster

▸ A 17″ color monitor

▸ A laser printer

▸ A 32 MB video/audio card

▸ High-quality speakers

▸ A 40x CD-RW player

▸ A scanner

▸ A color printer

Although it has not become much of a factor in cybercovens, a CD drive provides background music during rituals and is a good way to install additional applications on your computer. It is tough to find a new computer without a CD drive already installed. The scanner is an optional tool that adds information, such as graphics, that you can share with your coven. Having a color printer is just plain marvelous. Many of these options cost less than $200 each. Often, they are discounted if you buy them as a package. Check your local computer stores, shop the Internet, or buy a local newspaper and browse the used computer equipment section. If all else fails, just bribe your friendly neighborhood geek with lunch or dinner and take him or her on a trip to the local hardware outlet.

If you do not have a computer, then I recommend spending as much as you can afford on a system. Computers are useful tools for many reasons other than connecting to the Internet. If you already have a computer, then getting online is easier than you think, and it can introduce you to a whole new world of fun.

Many bargains in used computer equipment are available if you are willing to do some research. A great online source for getting excellent deals on computer parts and equipment can be found at http://www.price-watch.com. Cnet (at http://www.cnet.com) is another great resource for getting up-to-date information and recommendations about computer equipment, as is eBay (http://www.eBay.com). It is worth mentioning, however, that you need a computer to get online, which makes shopping for a computer online useless if you do not have access yet. Borrow a friend's computer, or surf the Web at the office, after hours. I have included a list of 800 numbers for some of the computer manufacturers in Appendix C, Pagan and Internet Resources.

If you spend a lot of time on the Internet, invest in a second telephone line specifically for the modem to avoid tying up your telephone. Cable or a DSL connection to the Internet do not require a second telephone line. Since these services are moderately to very expensive, depending on where you live, shop around for the best deal. Unfortunately, these services are not available everywhere.

SOFTWARE REQUIREMENTS—THE NITTY GRITTY

Once you have your basic equipment in place, simply add the proper software to access the Internet. Good software can make your online journey much easier. If you buy a new computer, it frequently comes bundled with a software package to get you online.

Remember that Cyberspace is not linear, but multidimensional; your movement from one Web site to another is not preplanned, and the variety of pages you can visit is infinite. Also, the World Wide Web is not the Internet, although the two are related. To reach the Web, you must connect to the Internet; but you can access the Internet and never go to the Web. The Internet is a network of computers that are constantly sharing information with one another. The WWW is a graphical environment that exists separate from, yet dependent on, the Internet.

Many books are available about how to use your computer, access the Internet, and navigate the Web. Use your local library as a resource rather than spending the money to buy a lot of computer books. Most computer information becomes outdated quickly. Local computer-user groups are also a source of helpful information.

ACCESS

Access to the Internet comes through an Internet Service Provider (ISP). The ISP provides you with a local telephone number that your modem dials to connect to the Internet. For your computer and the server computer on the other side of the connection, this action is just like you calling another person. The faster your modem, the faster the two computers communicate with each other. Another analogy is that connecting to the Internet through an ISP is like the old days, when you had to call your local switchboard operator to place a call.

National ISPs include companies like Earthlink, America Online (AOL), and AT&T Worldnet; many local services offer the same access. What's the difference between national and local? If you are new to the Internet, then a national service might be easier to understand and access. Bigger services tend to have simple-to-use installation software and a 24-hour customer service telephone line that you can call for help. But these services add up quickly, so pay attention to how much you spend, especially with per-minute charges above a flat monthly fee.

Prices vary, but about $20 per month will get you unlimited Internet access. Sometimes the service includes a free Web site in the price. ISPs are still a relatively new market, and it pays to shop around to compare prices and services. If you cannot decide between several providers, sign up with one and then switch after a few months. Keep switching until you find a service that is reliable and inexpensive. The downside is that your e-mail address changes every time you switch, making it difficult for people to stay in touch with you. Unlike the U.S. Postal Service, ISPs do not offer e-mail forwarding. A Web site at http://webisplist.internetlist.com/ provides a list of ISPs, divided by region.

Most ISPs provide you with software programs through which to access the Internet; the quality of these varies widely. Some systems allow you to use a netname, which is similar to a nickname, as part of your e-mail address; others always display your legal name. Finally, if you teach, attend, or work at a university, you are probably eligible for a free e-mail account. Some corporations also offer free Internet accounts to employees. If you choose to explore the Internet through an office computer, please make sure you check your company policy on Internet use. At the least, your company might frown on personal Internet use; at worst, it could be grounds for your dismissal.

An alternative to traditional modems and ISP services are cable and DSL (Digital Subscriber Line). DSL uses the same copper wires as telephone lines to deliver high-speed Internet access to almost any location. DSL and cable can operate at speeds as high as 1.5MBps and above. Variables include your location, line quality, and the distance to your local telephone company's central office (CO). Most existing home telephone wiring can handle a DSL line.

A fairly new option for home use is cable Internet service. This service offers you Internet access that is connected to your computer by a cable (similar to the kind of cable that provides good television reception) instead of a telephone line. Cable speeds are also very fast, as much as 2 MBps, but depend on line quality, the number of users on the line, your computer's performance and configuration, and overall network traffic. Generally speaking, if you have television cable, you can add on Internet cable.

Each service may or may not be available in your area, and it can be expensive. DSL installation may cost $100 or more, and service fees could be as much as $100 per month. Cable is less expensive per month (generally around $50), but the installation fee can still be about $150. These fees vary by provider and location. Companies occasionally offer promotions that significantly lower your initial costs.

When my partner and I switched to DSL, we ended up not spending any more money than we had with our (much slower) modems. Here is a quick look at our cost analysis:

Modem	DSL
2 telephone lines. . . . $40	1 telephone line . . . $20
2 ISPs. $40	DSL service $60
speed . . . . 33 KBps	speed. . . . . 1 MBps

When we moved, we lost access to DSL, but switched to cable; so now, we pay only $40 a month and have even faster (2 MBps) download times. Do your homework before you invest.

BROWSING

Once you have connected to the Internet, you need a browser, which is a program that allows you to move around on the Web, find sites, and collect data. Such a wealth of information is available to pagans online that I think some kind of Web browser is necessary, rather than optional. AvatarSearch (http://www.avatarsearch.com) is a pagan-oriented search engine that is particularly good for eliminating the hundreds of porn sites that come up if you type the word "goddess" into your search engine.

A browser-only interface is Opera, a small and extremely powerful program available at: www.operasoftware.com. This program is the only one that costs money; although you can download it for a 30-day free trial before you buy. You also need an e-mail reader program. Netscape Communicator and Microsoft Internet Explorer can receive and sort large numbers of messages as well as provide a browser. Each is free to download from the Web and provides many functions in one package. However, if you are not interested in the Web at all, Eudora offers a version of its e-mail reader, called Eudora Lite, which is available as a free download at http://www.eudora.com.

The best way to practice browsing (also known as surfing) the Web, is to look for information about subjects that interest you. But, because the Web is not indexed in any standard manner, finding information can be difficult. Search engines are popular tools for locating Web sites on certain subjects, but they often return thousands of results. They operate by logging the words from the Web sites they find in their databases. Because some search engines have logged the words from over 500 million documents, results can be overwhelming. Without a clear search strategy, using a search engine is like wandering aimlessly in the stacks of the Library of Congress trying to find a particular book.

Successful searching involves two steps:

▸ You must understand how to prepare your search. You must identify the main concepts in your topic and determine any synonyms, alternate spellings, or variant word forms for the concepts;

▸ You must know how to use the various search tools available on the Internet. For example, search engines, such as AltaVista (http://www.altavista.com), are different than subject directories, such as Yahoo! (http://www.yahoo.com). Both can vary greatly in size, accuracy, features, and flexibility.

Subject directories are often called subject trees because they start with a few main categories and then branch out into subcategories, topics, and subtopics. As an example, if you are interested in finding information on Paganism through Yahoo!:

1. At the top directory level, choose **Society & Culture**

2. At the next directory level, choose **Religion**

3. At the third directory level, choose **Faith & Practices**

4. At the final directory level, choose **Paganism**

Because humans organize the Web sites in subject directories, they are most useful for finding information on a topic when you don't have a precise word for which to search. Many large directories include a keyword search option, which usually eliminates the need to work through numerous levels of topics and subtopics. Directories cover only a small fraction of the pages available on the Web, so they are most effective for finding general information on popular or scholarly subjects. If you are looking for something specific, use a search engine.

Search engines differ from subject directories. With a search engine, you type keywords related to a topic into a text box, then click the **Search** or **Go** button. The search engine scans its database and returns a file with links to Web sites that contain the word or words that you specify. Because these databases are so large, search engines often return thousands of results. Since topic words can appear several times on one page, or on many pages throughout a single Web site, the results may contain thousands of duplicate listings. Without search strategies or techniques, finding what you need can be like looking for a needle in a haystack.

To use search engines effectively, you refine the search to push the most relevant pages to the top of the results list. A list of useful techniques follows. I recommend that you choose a search engine and try these techniques to learn how they differ.

1. When conducting a search, break down the topic into key concepts. Instead of a broad search for "Paganism," narrow your criteria by typing `pagan parenting`.

2. Use Boolean operands. Boolean operands are words such as "and" and "or" that connect words. By using these operands, you tell the search engine to retrieve Web sites that contain all or some of the keywords. For instance, if you type in pagan and parenting and advice:

▸ The search engine will not return pages with just the word "pagan";

▸ Neither will it return pages with the words "pagan" and "parenting";

▸ The search engine will only return pages where the words "pagan," "parenting," and "advice" all appear somewhere on the page.

Thus, "and" helps to narrow your search results because it limits results to pages where all the keywords appear. On the other hand, "or" widens the scope so that the search engine only returns pages where the words "pagan," or "parenting," or "advice" appear somewhere on the page. The use of the phrase "X and Y and not Z" will eliminate pages from the same search; "pagan and parenting and not advice" will produce yet another list of Web sites.

Some search engines accept the plus (+) and minus (-) signs as alternatives:

▸ The plus sign (+) is the equivalent of "and";

▸ The minus sign (-) is the equivalent of "and not";

▸ Do not leave a space between the plus (+) or minus (-) sign and the keyword(s).

Phrase searching is another technique for finding information on the Web. If you surround a group of words with double quotes (" "), you tell the search engine to retrieve only documents in which those words appear side by side. Phrase searching is a powerful technique for significantly narrowing your search results, and you should use it as often as possible.

In general, most search engines interpret lowercase letters as both uppercase and lowercase letters. Similarly, most search engines interpret singular keywords as singular or plural.

▸ If you want uppercase and lowercase occurrences returned, type your keywords in all lowercase letters.

▸ If you want to limit results to words with initial capital letters, such as Starhawk, or all uppercase letters, type your keywords that way.

▸ If you want to return both singular and plural forms, make your keyword singular.

▸ If you want plural forms only, make your keyword plural.

A few search engines support truncation or wildcard features that allow variations in spelling or word forms. The asterisk (*) is a wildcard symbol that tells the search engine to return alternate spellings for a word at the point that the asterisk appears.

▸ "capital*" returns Web sites with "capital," "capitals," "capitalize," and "capitalization."

At first, you might have trouble navigating from Web site to Web site and locating what you want. If you get lost, simply click the **Back** button in the upper left corner of your screen until you regain your bearings. Sometimes, your seemingly simple search returns hundreds of thousands of sites. Try different search engines, alter your search syntax, or try using different search words altogether. Each action will change your search significantly. Learning to use the Internet properly takes time, practice, and patience.

You could discover that you require different software to accommodate your needs, or that physical problems make it difficult for you to use your computer for long periods of time. Don't worry—a little ingenuity and per-haps some advice from the local computer store can help you to work around problems. For instance, you might need a special keyboard, mouse, or chair to make computer use comfortable for you. Hunt around; the mar-ket carries a wide range of adaptive products.

You may need to purify and consecrate your computer as a ritual tool in order to use it comfortably. One way to cleanse and consecrate your computer and related devices as magickal tools is to do a simple ritual of intent.

/Ritual of Intent for Computer Components/
Lay out the components neatly, cords coiled and vents vacuumed. Have your chalice (filled with fresh water), salt, incense, and a candle ready. Cast the circle by facing each direction in turn and saying:

Creatures of the East, Elementals of Air,
I ask your presence, your boon of Inspiration.
Come in peace, so mote it be!
Creatures of the South, Elementals of Fire,
I ask your presence, your boon of Energy.
Come in peace, so mote it be!
Creatures of the West, Elementals of Water,
I ask your presence, your boon of Emotion.
Come in peace, so mote it be!
Creatures of the North, Elementals of Earth,
I ask your presence, your boon of Wisdom.
Come in peace, so mote it be!

As you walk the circle, visualize a line of blue or white light tracing your path. Stand in the center and raise your arms overhead. Visualize a sphere of light forming below and above you, enclosing you and your tools completely. Say:

Gracious Lady,
Generous Lord,
Thou who wast here before all,
Thou who will endure after all,
Bless my circle with Thy presence.
Come in peace, so mote it be!

Light the incense, and then the candle. Quickly move each component through the incense, saying:

Computer creature, be my friend.
Protect my data, be free from harm.
Computer creature, hear my plea.
The strength of Air binds this charm.

Then (quickly) move each component through the candle's flame, saying:

Computer creature, be my friend.
Protect my data, be free from harm.
Computer creature, hear my plea.
The strength of Fire binds this charm.

Mix a little salt into the water and stir to dissolve. Mark each of the computer components with a star inside a circle, saying:

Computer creature, be my friend.
Protect my data, be free from harm.
Computer creature, hear my plea.
The strength of Earth and Water binds this charm.

If you are moved to add other words, please do so. You may also want to spend a little time reflecting on the wondrous power of the computer, and how much good it does in your life.

When you are ready to finish, move to the center of your sphere and say:

Sweetest Lady,
Noblest Lord,
I thank you for Your presence
In my circle tonight.
Go if you must, stay if you will.
Blessed Be!

Open the circle by facing each direction in turn, saying:

Creatures of the North, Elementals of Earth,
I thank you for your presence, your boon of Wisdom.
Go in peace, Blessed Be!
Creatures of the West, Elementals of Water.
I thank you for your presence, your boon of Emotion.
Go in peace, Blessed Be!
Creatures of the South, Elementals of Fire.
I thank you for your presence, your boon of Energy.
Go in peace, Blessed Be!
Creatures of the East, Elementals of Air,
I thank you for your presence, your boon of Inspiration.
Go in peace, Blessed Be!

As you walk the circle, visualize the blue or white light disappearing.
    It is done.

Above all, stick with it. Soon you will feel as comfortable on the Internet as in front of your altar.
    Now that you are connected, let's explore!

3: Exploring and Communicating

Once you connect your computer to the Internet and are comfortable exploring the Web a bit, you can explore other aspects of your newest occult tool. You can do many fun things online, like send e-mail, visit FTP sites, and converse in Usenet groups. The two tools that will radically increase your knowledge and ability to interact with other like-minded pagans online are the (e-mail) list and IRC (Internet Relay Chat).

COMMUNICATION BASICS: E-MAIL, FTP SITES, AND USENET NEWSGROUPS

E-mail is the basic medium of communication online. Some companies offer free e-mail accounts (like Juno and Yahoo!), and most ISPs include e-mail with the basic package. Through e-mail, you can send messages to several people at once or forward a message sent to you on to someone else. Mailing lists, which I explain later in detail, are a form of e-mail. You can send text messages, photographs, and links to Web sites. Or you can attach documents from your word processor or a spreadsheet package, or even send a recorded sound message. Most e-mail programs, such as Netscape, allow you to read and write these messages when not connected to the Internet, which may save you telephone charges. Unfortunately, the mechanics of sending e-mail—which buttons to click or how to attach a photograph—vary with every e-mail package. Consult your user manuals. Also, use the help menus and read directions.

/Basic FTP Techniques:

File transfer protocol (FTP) allows you to move any kind of data to and from thousands of other computers. Occult Web sites can provide interesting and unusual information, including spell books, graphics, poetry, and ancient literature. FTP sites are an excellent way to find guides to software and hardware, newsletters, archives, and other information on a wide range of topics. If you find information you want to download:

1. Click on the **Link** to the information.

2. You will see a specific file or a list of files in a directory.

3. Click on the **File Name** of the information you want to download.

4. You will see a dialog box that indicates that the file is about to be downloaded. You can usually choose where on your computer to store the file. I recommend that you create a specific directory for such downloads, perhaps "My Downloaded Files."

5. Click **OK** and the file is downloaded to your computer.

/Newsgroups:

Newsgroups (Usenet) started when a few people at a few campuses wanted to share information by posting it online so that anyone could read it and respond. Legend has it that, maybe ten years ago, you could read every message in every group over a single cup of coffee. Today there are tens of thousands of newsgroups on every topic imaginable. Each Internet service decides how many of the newsgroups to make available; but most services provide at least a thousand for their clients to access.

Through the use of newsgroups, terms such as "spamming" and "flaming" originated (see Glossary, page 147). As an example, if you spam a group, you are likely to get flamed. Reading newsgroups can be fascinating and useful. Depending on the size of the group, asking a question can elicit replies from all over the world. Some groups focus on news, others on discussion. The overall topic may be broad, narrow, or random. Newsgroups sometimes have a moderator to filter out inappropriate postings. Internet manners, or "Netiquette," suggest that you read posts for a few days before sending your first message. The practice of reading without responding is called "lurking" and gives you a chance to determine the atmosphere of that particular group before participating. You should also find and read the Frequently Asked Questions (FAQ) list, which will answer most of your questions. A great, general FAQ list for Usenet groups is available at http://www.faqs.org/faqs/usenet/what-is/part1/.

LISTS

The Internet provides thousands of discussion groups by allowing users to place themselves on electronic mailing lists via e-mail. Similar to a Usenet group, a software program, such as LISTSERV(R), LISTPROC, or Majordomo creates a list. This list automatically distributes an e-mail message from one member of a list to all other members on that list. So, if you want to tell a large group of people about an upcoming meeting, you send a message to a single address and the entire group receives it. Thousands of lists in the form of digests, electronic journals, and discussion groups are available online.

When you subscribe to a list, your name and e-mail address are automatically added to the list. You usually receive a welcome e-mail telling you about the list. From that time on, you will receive all e-mail, called postings, that other members send to the list. You can simply follow the discussions or participate. If you respond, you can send your response to the list so that all members receive it, or to an individual on the list. You can sign off or unsubscribe from a list at any time. Sometimes, you can get a roster of all the members of a list and their e-mail addresses, depending on the list's level of security.

You can create a mailing list of your own within any one of the larger e-mail programs, but I do not recommend it if your group has more than ten people. Over the course of a year, my first cybercoven's list had more than 15,000 messages posted, and our coven averaged twenty-eight members. This amount of online traffic represents a lot of volume for one person to handle, but is not a problem for a program. My current cybercoven generally has 1,000 messages a month, with eighteen members.

Several companies will host your list for free, including Yahoo! Groups (http://www.yahoogroups.com). Others charge a small fee ranging from $10 to $50 per month, depending on the list size and number of members. The reason Yahoo! Groups is free is that every single message contains an advertisement. The ads can be annoying, but my cyber list members have not been irritated enough to pay money each month to have them removed and, sometimes, they can be hilarious. A conversation regarding honesty and communication accompanied by ads for an online mental health service was particularly funny.

List features vary, but generally include easy methods to add and delete members, change addresses, and monitor posts. I am the manager for several lists through Yahoo! Groups and participate in about thirty others. My coven has a list for the class we teach—not only for the lessons and their response, but also for meditations and insights about our lives. This blending works well, because we believe that the lessons we teach are not separate from life. Yahoo! Groups has a calendar function that I use to send out reminders about our online classes, rituals, and meetings. An archive stores every single e-mail sent out on the list, which is a blessing if you misplace a lesson or your computer crashes.

Using Yahoo! Groups and JaguarMoon coven as an example, follow these steps to create a list using a service:

1. Type `http://www.yahoogroups.com` in the **Go To** line of your browser.

2. Look for a phrase like **Start a Group Now.**

3. Click on this phrase with your mouse.

4. Type in a name for your group, for example `JaguarMoonCoven`.

5. Type in the e-mail address to which members will post messages, for example `JaguarMoonCoven@yahoogroups.com`.

6. Type in a description for your group, for example: `A cyber coven in the ShadowMoon Tradition`.

7. Choose the primary language of the list by clicking on it.

The site will then ask you what setup you wish your list to have.

▶ Choose **Listed** if you want any visitor to Yahoo! Groups to find your group in the directory.

▶ Choose **Unlisted** if you want to prevent public viewing.

▶ Choose **Open Membership** if you want to allow any interested party to join your group.

▶ Choose **Restricted** if you want the list owner or moderator to approve all members before joining.

▸ Choose **Private Membership** if you want to allow only those people directly invited to join.

When choosing a moderation status, remember that, if you are choosing and approving members, it is less likely that you will need to moderate their posts, because you will know who the members are. But in an open list, placing new members on moderated status to begin with can prevent inappropriate posts from bothering your list.

You will need to locate your group in the appropriate category. To do this, think about where your group is best categorized, because it will make a difference. If you aren't sure, browse through groups that are similar to your own, and see where they are located.

You are then asked to invite a list of people to join (this is optional), choose how they will be notified, and enter an introductory message. Send it and you are done!

Creating a group automatically makes you the list manager. To change some or all of your list's policies, or to make another person the list owner and manager:

1. Log into your Yahoo! Groups.

2. Click on the **Manage** tab. You can change a variety of options:

   ▸ *Subscriptions*: Anyone may join the group or approval is required to join the group. The list manager is required to approve each new member before he or she can join the group.

   ▸ *Postings*: This determines who can post to your list.

   ▸ Click **Only Members** if you want to prevent nonmembers from posting and keep "spammers" from posting irrelevant material and polluting your group with their nonsense.

   ▸ Click **Only The Group Moderator** if you want to make the group essentially an announcement list, suitable for solicited business promotional announcements.

   ▸ Click **Anyone** if you want to gather feedback from anyone on a particular topic, without expecting others to follow future discussions.

▸ Click **Nobody** if you want to close down all discussion.

▸ *Moderation*: Posted messages are first forwarded to the moderator for approval. Choose this option if you want to approve every posting before it is distributed. This option can be a lot of work, but provides value to group members. You can share the work of moderation with another member. Posted messages can also be distributed directly to the group.

▸ *Web View*: Messages can be read on the Web by anyone or can be read on the Web by members only. An up-to-date browser that supports cookies is required to access password-protected Web archives.

▸ *Member Listing*: The group has no member listing, not even for members, which provides maximum privacy. Alternatively, the group has a member listing that its members can see, in which case, only members can see the information other members have chosen to share.

▸ *Reply-To*: You can reply directly to the sender of a message, or you can send responses to the entire group. This second choice is more convenient for discussion groups of limited size.

Yahoo! Groups has a constantly updated and informative help page at http://www.yahoogroups.com/help/groups.

INTERNET RELAY CHAT (IRC)

For an even more powerful tool for communicating online, you can download a copy of Internet Relay Chat (IRC). IRC is a program that allows you to create a channel that is available twenty-four hours a day. It is a real-time forum for discussions, meetings, classes, and rituals. This channel is vital, because all of your group rituals will be held there.

To download IRC, go to the mIRC homepage at: http://www.mirc.com and look for the version most compatible for your computer type. To install, follow the instructions on the Web site. To load and use the most basic applications of this program:

1. Connect to the Internet using your cable, dial-up, DSL, or other ISP service.

2. Double-click on the **mIRC Icon** on your desktop.

3. You will see a dialogue box labeled *About mIRC* with a picture of the author in the corner opens.

4. Two choices appear: *Introduction* and *How to Register*. For now, ignore these choices and register later.

5. Click on the **X** in the upper-right corner of this dialogue box.

6. You will see another dialogue box labeled *mIRC Options*.

7. Type your name, your e-mail address, your preferred nick-name, and an alternative, if you have one. (Note: IRC names cannot include punctuation marks, so Ma'at becomes Maat.)

8. Within the same dialogue box, click on the **Down Arrow** to the right of the server listed.

9. Click on the **Server** you want to use. The most popular and oldest are: UnderNet, DalNet, IRCNet, and EFNet.

10. Choose a specific location to dial into, or click on the **Random Server** option. Remember which server you are on so you can tell someone where to meet you.

11. Click on the **Connect To Server** button.

12. You will see a screen labeled *Status*, with a single line of text that reads: *connecting to irc.servername.net (####)*. This text indicates that your computer is trying to reach the server that you requested at the port (####), or to an automatically programmed address. Many lines of text will scroll on your screen.

13. You will see another dialogue box that asks you which channel you would like to join. Channels are where the conversation is actually held.

14. Choose an existing channel by clicking on the **Channel Name.**

15. To create your own channel, type #nameofchannel in the text box.

If you are new to IRC, joining the beginner channels is an excellent way to practice and learn. Generally moderated by an experienced user, users on the beginner channels can help you learn to maximize your experience on IRC.

Gather some friends and go chat in a channel. Read the help files and try out all the commands. IRC is a user-friendly application and works on any operating system. You can register channels that you frequently use for regular meetings or rituals; or create a single-use channel that is closed after all the members leave. Using IRC is easy to learn and, since so many people use it, help for advanced applications is easy to find.

The downside of holding meetings on IRC is that it can be difficult to ensure privacy, because any stranger can join channels, even if you specify that it is a private channel. You can use various tricks to lessen the likelihood of intruders.

To keep your privacy, I recommend registering the channel and users' nicknames with the server that you specify. By registering, you can create "auto-ops" for certain people. Auto-ops is a useful feature, because channel operators (Ops) can set channel topics, invite people to join, or oust people who do not belong.

Note: The first person who joins a nonregistered channel becomes an Op, and he or she has the same privileges as a registered operator. Also, you must visit your registered channel on a regular basis, at least every week on some networks, to keep the registration active.

When in IRC, remember three things:

▸ All IRC commands follow the forward slash (/).

▸ All channel names begin with a pound sign (#).

▸ IRC is not case-sensitive, so you can use uppercase, lowercase, or mixed-case words in your commands.

Not all servers allow channel and nickname registration, and exactly how the registration happens varies. Use these general guideline to register:

▸ To register the nickname that you currently use, type `/nickserv register <nickname> <e-mail address> <password>`.

▸ An alternative is to type `/ns register <e-mail address> <password>`.

▸ On any server, if you need help, type `/msg NickServ help register` to bring up information on how to register your nickname.

▸ To learn how nickserv and chanserv work on your server, type `/nickserv HELP` or `/chanserv HELP`.

After you register your nickname, you must identify yourself to the server each time you log on.

1. Click on the **File** menu, then click on **Options**.

2. Click on **Perform in IRC**.

3. In the text box, type `/msg nickserv IDENTIFY <password>`.

4. To register your channel, type `/chanserv REGISTER <channel> <password> <description>`.

5. An alternative is to type `/cs REGISTER <channel> <password> <description>`.

6. On any server, if you need help, type `/msg chanserv help register` to bring up information about how to register your channel.

If you want to join your channel permanently every time you log on to IRC:

1. Click on the **File** menu and click on **Options**

2. Click on **Perform**

3. In the text box, type `/join #nameofyourchannel`.

When you hold meetings, set the channel to a mode in which outsiders cannot see the channel. To do this, have an operator type `/mode #channelname +s`. In this mode, people who want to join the channel must know that it exists; therefore, this action eliminates visitors who are just surfing. Some people troll the Internet looking to cause trouble, which brings me to my next topic.

ONLINE PRIVACY AND ANONYMITY

One of the greatest advantages to joining a cybercoven is that your participation is not limited by geography. If you have access to the Internet, you can be a member. The second greatest benefit is that you can remain "in the broom closet" within your physical world, which is a big concern for many witches. Online, you are known by several names:

▸ The name you give your e-mail program to display in all messages you send:

> To: John TallBerries (johnnyt@skienet.net)
> From: "Lisa" lisa@cybercoven.org
> Date: January 13, 2000
> Subject: Naming and Numerology

▸ The craft name or other name by which you choose to be known by:

> John TB,
> I was just wondering where you chose your name from. Is it mythological?
> In love,
> Ma'at

▸ Your ritual name, which may or may not be your craft name, which you choose during your initiation, and which may or may not be a name you choose to make public online:

> *Welcome to sandnet.org*
> `/join #thetemple`
> *Maat has joined #thetemple*
> *Maat changes her name to Lady_Maat*
> *Lady_Maat>* `Greetings everyone!`

Your computer and ISP also have assigned you names that indicate such interesting pieces of information as your location on the Internet and how to find you to send you your mail.

In a sense, you can go online and assume a role along with your online name. Although members must inform several people—the High Priest or Priestess, the ISP billing department—of their true information, everyone else only knows what you reveal about yourself. Other revealing pieces of information are also shielded, such as where you live and what you do for a living. In most cases, you are probably the only one who will inform others about what you do in the physical world.

For example, we had a member in our first year who worked in a government-sponsored facility with a top-level security clearance. This person was concerned about having her employers find out about her identity as a witch. She made it clear to the High Priestess that her personal information was not to be revealed. To the coven, she simply stated that she did not like to discuss her work life within the coven. We all respected her need for anonymity. Although it was odd not knowing what state she lived in, we all got to know her and, after a year or so, she trusted the group enough to share stories about that part of her life.

Despite what the U.S. Constitution states, religious freedom is not something to take for granted and we must all respect the need for privacy. As a member of a cybercoven, you have some control over of how separate the various facets of your life will be. Even if you enter "none" under religion on your employment form and never wear any religious jewelry, you may find that it is okay if a few people at work know of some of your beliefs. You may be surprised where you find other devoted pagans. You may eventually decide that it is an acceptable risk to let other coven members know some details of your "physical" life. Then again, disclosure can present an unacceptable risk if you live in an environment that is hostile to Paganism, such as the Bible Belt. Joining a cybercoven may be a safe way for you to explore pagan spirituality.

Now that you know something about the options available to you in Cyberspace, think about what you can do with them. If you just discovered Paganism recently, welcome to a rich and diverse community and abundant resources. Here you can find teachers, news about upcoming events, details on religious beliefs other than your own, and a feel for what different branches of Paganism are like. If you have practiced Paganism for a long time, you can share your experience with newcomers, talk with experienced members of many systems, and further your knowledge of the Craft. You can find whatever you need here; you just have to know where to look.

/part_two/

Interactions

4: Finding and Joining Other Pagans

Almost daily, I see messages posted to lists and newsgroups along the lines of: "Help! I want to learn more about such-and-such, and there is no one in my area to teach me! Anybody with information about joining a coven please e-mail me." Every week, people ask me about joining my coven and I can empathize with their frustration and desire. Prior to joining ShadowMoon, I went through periods when I wished I had someone with whom to share knowledge and magick. I especially wanted to surround myself with people who were experienced pagans.

WHAT IS A COVEN (AND WHAT IS IT NOT)?

A coven may also be called a temple, or grove, or circle. If a coven has the right blend of people, it is a marvelous entity. According to Starhawk, "The coven is a Witch's support group, consciousness-raising group, psychic study center, clergy-training program, College of Mysteries, surrogate clan, and religious congregation all rolled into one."[1] Each coven has its own personality, woven from each individual's contributions. Ideally, a coven is the training ground in which each member develops his or her pagan potential, not just a group of people who get together to work magick.

Then again, while I sometimes found it difficult, practicing as a solitary provided a great opportunity to grow without externally imposed limitations. I needed to learn to rely on my own perceptions and abilities, instead of looking to somebody else to spoon-feed me enlightenment like a baby.

When I was ready to find a coven, I had already spent a long time study-ing in a semi-structured learning environment.

A coven is a unique entity in that each member must understand that self-confidence, sincerity, ambition, and absolute honesty are critical char-acteristics of a dedicated witch. While even the most traditional coven allows for some differences in belief, personal agendas that do not serve the interests of the entire group have no place there. It may take many years to find the right coven. Some people may work with several covens of various traditions before finding one that feels right. You must trust the gods and goddesses to lead you to the right teacher when the time is right. And even then, there are no guarantees that the student-teacher relation-ship will last forever. Most people have many teachers over time and glean valuable information from all. Your job as a member or prospective mem-ber of a coven is to be honest with yourself and the coven leaders, and to trust the gods and goddesses to guide you along the right path.

There are several excellent reasons for joining a magickal group:

Increased strength. Teamwork can accomplish amazing things. In mag-ick, a coven usually has a lot more strength than a solitary practitioner. In a close-knit group, all share responsibility for the outcome.

Companionship. In a good coven, your brothers and sisters are people you can trust with your personal issues; they can offer advice and support. When you have five or six people you can call on if your familiar is ill, you know where your emotional support is coming from, or whom to call to help you mourn, if things turn out that way.

Wider range of information. You can find more opportunities to learn in a coven. Everyone has a different learning and teaching style, and each witch leads differently. In a group of people where ideas are exchanged freely, a lot of learning occurs. Other groups exist online; but the intimacy of a coven seems to create a fertile environment for spiritual growth.

That said, remember that a coven is not:

Group therapy. Some people who are drawn to the occult are damaged or dysfunctional in some way, yet still manage to cope without the constant need to lean on others. Wicca is not a 12-step program; if someone has a problem with addictive or compulsive behavior, he or she should work it out in therapy; not in the coven. Yes, crises arise now and then, but repeatedly wasting group time on an individual's personal issues only weakens the group. Support groups help people work through problems that they find difficult or impossible to handle alone.

A nuclear family. A good magickal group feels like a healthy family. Just remember that the leader is the group's mentor and inspiration, not its parent. The leader must never project the image of Mommy; nor should the students assume childlike roles. If you observe that these roles occur in your coven, think carefully about why.

A sexual playground. Within the intimacy of a well-running magickal group, sexual activity may become a temptation, but it is almost never a good idea. Some Neopagan groups advocate polyamorous lifestyles. Yet, in my experience, a coven that prays and lays together, does not stay together. If two members who are lovers break up, then the circle witnesses the breakup in uncomfortable detail. The negative energy from such situations can bleed into the coven's collective energy and destroy it.

A party. All acts of love and pleasure may be Her rituals, but please remember that balance is vital to the health of a coven. If you want to imbibe or indulge on ritual occasions, go ahead. But coming to rituals drunk or high is a waste of your time and can destroy the energy in your group. Likewise, random interruptions or pure silliness is distracting, annoying, and sometimes just plain rude. After the ritual solemnity is a better time to relax, let off steam, and get loose.

A social club. More specifically, your coven should not be your only outlet for social activity. You may find it worthwhile to relate to others without revealing your spiritual focus, because, sooner or later, the other members of your group will move on, leaving you to wonder why they never call you anymore. Your activities in the real world are just as vital to your health and emotional well-being as any you may pursue in the pagan community.

A dictatorship. If the group's leader is manipulative, makes arbitrary rulings, withholds approval, obsesses over enemies, or otherwise behaves irrationally, he or she does not deserve your love and respect.

When witches practiced in secret, potential members were limited to finding other like-minded people within their community. Back then, seekers hoped that the others held the same or similar viewpoints, had knowledge to share, and were reasonable, sane, and balanced. With money or time, our travel options expanded; but geography remained a limiting factor. Cybercovens are free of this constraint and have proliferated on the Internet. But how can you tell which coven best meets your needs?

WHAT THINGS DO YOU WANT IN AND FROM A COVEN?

In deciding whether to join a coven, consider what you want from your magickal group. Write down a detailed list in your journal, if you want. Meditate on it. Incorporate that list into a ritual, if you feel guided.

There are good reasons to join a coven and there are bad reasons. Here are a couple of the worst reasons for joining a coven:

▸ Are you just dying to get into a circle where you can do some real fancy magick? Perhaps you just want to conjure spirits or win the lottery; or maybe you just want to do Magick with a capital M—just because! This is not a good idea. In covens as well as in solitary magick, "fireworks" spells are just a waste of energy. They drain the practitioner and serve no purpose other than ego-boosting theatrics. Spell work can also be addictive to some kinds of people. Do not put yourself in the position of always searching for the next "fix." Real power lies in being able to accomplish your goals in life, not trying to impress others with your abilities.

▸ Related to creating fireworks is joining a coven to show off your personal knowledge and skills. You have read every book and done every spell and ritual successfully. You walk lightly upon the Earth and live an entirely seasonal lifestyle in which plants and animals literally come and offer themselves up as sacrifice so you can eat. But, since living this life is lonely and you have no witnesses, you have decided that you want to join a coven and get external validation. Definitely climb down off the mountain and open your eyes. Realize that there will always be more to learn and teachers to provide that knowledge. Sharing knowledge is a generous act; but, if you want more than a thank-you, then do your coven a favor and go back to being a solitary practitioner.

▸ At the other extreme, are you feeling sheepish? Sometimes people flock to covens because they need someone to tell them what to believe and to do. These flocks do not trust their own observations, or they have become so accustomed to having others think for them that they have no sense of self. It is one thing to be new enough that you do not know where to begin or what questions to ask. I believe we all need at least some structure in our lives. Be cautious, however, with people who want to be told what to do. If you are looking for a High Priestess to tell you how to live, perhaps you need to stop and consider some alternatives. Spend the time to focus on your beliefs, piece them together through your observations from life, and strengthen your faith. In the end, you will know what you believe and why. When you accept a leader without testing what he or she says against what you've already learned, you give away too much of your own power. The more you learn, the more your leader has to account for.

Most likely, you will select a coven that matches your beliefs about the universe and Divinity, although it may not be possible to find and unite several people who worship your God or Goddess, especially if you resonate with a specific deity. In a coven based on a pantheon, or dedicated to specific deities, everyone in the group relies on the same pantheon. Covens of this type might include those dedicated to Celtic, Egyptian, Greek, Roman, or Hindu pantheons. A unified coven has an advantage when it comes to appealing to their deities for favors and gathering information about them. In ShadowMoon, for example, we are dedicated to the goddess Hecate and the god Herne, although I am personally dedicated to Diana and Mercury. In group rituals, I call upon Hecate and Herne, but, when practicing as a solitary, I call upon my personal deities.

Consider your personal preference if you choose a group based on the way in which it's organized:

▸ You may learn better from a lecture than from open discussions;

▸ Perhaps you prefer to read information, instead of listening;

▸ Look at the coven leaders' authority. Are you comfortable with how they wield their power?

▶ The level of commitment required from coven members should be agreeable to you;

▶ If you do not like the way the High Priest/ess is running things, you probably will not want to stick around for very long.

In an eclectic group, consider the foundation of beliefs. They do not have to be the same as yours, but they should not conflict either. For example, if you do not feel comfortable worshiping a male deity in any aspect, then you might prefer a Dianic coven, which is devoted to female deities exclusively. Things to compare include:

▶ Is your vision of Divinity compatible with the others in the coven?

▶ Do you celebrate similar Sabbats?

▶ Do your ethics agree enough to engage in magick without creating conflict?

▶ Does the group call circles and quarters in a way you find understandable and meaningful?

Also, ask several questions about any group—cyber or physical—with which you might want to work:

How long has the coven been in existence? This question is especially useful for cybercovens. As far as my research shows, few cyber magick groups exist longer than three years. Although there were cyber-workings as early as the mid-1980s, formal groups were extremely rare, and very few have remained intact over the years. If you read about a coven that claims to have been around for more than five years, check carefully: it might just be moving from the physical to the cyber realm. Newly forming covens can be exciting, but they also face unique challenges and frequently fail. Covens that have been around for a while tend to be more stable, but if you feel changes are necessary, you might leave before change is made.

How big is the coven? Too big to give each member a chance to feel as if he or she is a vital part of the group? Or, too small to spread responsibilities evenly and thinly? There is no perfect number, because the coven's focus influences its size. For example, in a teaching coven, the ratio of teachers to students really should not be more than one to four. Even that is a stretch; one to three is much better. If the coven only meets to do ritual, however, then the size can be much larger.

Who are the coven members? Are they of mixed gender, ethnicity, socio-economic backgrounds? Online, a diverse mix of people is easy to achieve. If the coven has a tradition based on one pantheon, the members may tend to be more of one type than another; but a variety of people should participate. If not, then you might want to ask why.

How are rituals done? Who chooses when, where, and what rituals are performed? Is that a structure comfortable for you? If want to join a coven to gain experience in writing and leading rituals, then being excluded from that process probably won't feel right.

How much time do you have, and how much are you willing to devote to being a part of a group? If you are already juggling too many demands on your time, then perhaps joining a coven is not a good idea right now. One of the benefits of cybercoven membership, however, is that you can gain companionship and knowledge at a pace that you choose. You still have to juggle your priorities and deal with conflicting demands, but cybercovens require no travel time.

Are they friendly in their dealings with you, but not over-enthusiastic about recruiting? Are they willing to talk with you about how they do things and what their expectations of you would be? Openness and good communication are essential characteristics of a healthy coven.

Do members seem to know what they are talking about and are they willing to share that knowledge? Are they emotionally and physically healthy? Do they feel like good people to be with? These questions are a little harder to answer when you are dealing with people online, but not impossible. Listen to your intuition.

Do members have fun? Do they laugh with one another? Is there laughter within rituals? If so, how is it handled? ShadowMoon, for example, has a resident Trickster. CoyoteMan's off-topic comments during rituals are sometimes inappropriate and occasionally the High Priestess tells him to stop; but we usually all laugh and move on. Believe me, if you've ever slipped and called the Elemental Watertowers, instead of Watchtowers, ritual can succumb to silliness in an instant.

Are there financial or political obligations? How are finances handled? Are there exceptions made? If so, how and to whom? Is the coven given regular reports? Has anyone left the coven or been banished? If so, can you speak with the former member? Speaking with former coven people may reveal more about the internal structure of the coven and why it is not as perfect as it seems on the surface. Ask why the person was banished. The answer to such a sticky question tells you much about the politics of the group, including whether you might find yourself similarly banished some day. No matter how much you want to work with a group of people, it is a waste of your time to join a group with which you are not comfortable. Instead, wait, practice on your own, and continue to look. It will be worth the wait.

WHAT THINGS SHOULD MAKE YOU LEAVE?

Some things that are negative in a coven will not be obvious at first. If you ever feel that something is wrong, or you do not like how a situation is being handled, then seriously consider whether you want to leave that coven. Use your common sense. People do make mistakes, but staying in a situation that does not feel good stunts your spiritual growth.

Here are some specific warning signs that you should leave a coven:

▶ Be wary of any coven leader who claims to have grasped the one and only truth, making all other beliefs and practices wrong. For a newcomer, it may feel as though a teacher is saying, "This is the only way to do that." Good teachers usually say, "This is the way that I was taught and it has worked well for me and for those I've taught." If a coven leader's philosophy and beliefs are narrow and one-sided, then back away.

▶ Does the coven leader(s) exercise too much authority? Do they try to control the personal lives of members? Does it seem as if there is a guru to whom everyone defers?

▶ Be wary of a group that keeps much of its doings secret. Never speak words in a ritual setting that you do not understand. Remember that you can always walk out of a ritual.

▶ Be wary of a coven leader who shows a lack of respect toward members. Related to this, shy away from any coven that has a lot of misunderstandings going on all the time. If there are constant flame wars and harsh words, followed quickly by apologies, beware. Flames and harsh words indicate a lot of disrespect. Occasional disagreements are okay and even healthy. But anger should not be a usual part of coven interaction.

▸ Are guilt-trips a part of coven discussions?

▸ Is the group overly concerned with how many members it has? Does it accept anyone who asks to join? Is the screening process essentially meaningless, or, even worse, skipped? Why is the group so concerned with having members? The answer to this question could show that attrition is a problem and no cure is available, or that "size matters" in an unhealthy way.

▸ Leave a coven that has no clear belief system or one that accepts everything from Sculderianism (a system that worships a dual deity in the form of Mulder and Scully from the TV show *The X-files*) to the Illuminati (a hidden magickal organization that theoretically runs the world) as valid belief systems.

▸ Be wary of any group that tries to alienate you from your family and friends. This is a dangerous warning signal of cults, which are not healthy groups for you or anyone else.

▸ Never join a coven in which it is unacceptable to excuse yourself from practices that make you uncomfortable.

▸ Do not remain in a coven where you cannot ask questions about anything magickal. You should be able to ask questions such as, "Why did you use sandalwood incense and not pine?"

▸ Do not join a coven that does not tell you what you will be learning or doing during rituals, nor one that provides no by-laws.

▸ Do not join a coven that uses drugs or alcohol in rituals.

▸ Do not join a coven that requires you to engage in sexual activities of any kind with any member. In a cybercoven, this is less likely to occur, but if sexually explicit flirtation is common in coven posts, use caution.

▸ Avoid covens that offer you a First Degree initiation after only a short time. (However, such an initiation might be in order if you have trained in other traditions or have been practicing as a solitary for more than decade.)

▸ Do not remain in a coven that requires that you Dedicate yourself fewer than two months after joining. Two months is barely enough time to get to know everyone, even in a cybercoven, where interaction can occur on a daily basis.

▸ Be wary of joining a coven or group that guarantees that you will be given an Initiation after a predetermined period of time. Initiation is too sacred an act to be defined by time.

▸ Do join a coven that follows the Wiccan Rede verbally as well as in practice.

▸ Avoid covens that participate in bindings, hexes, or witch wars. Do they require a physical link from you before you join? Do you ever wonder whether their magick will be used against you? Have you ever been told that if you leave you will suffer reprisal?

▸ Avoid a coven in which the members are unhappy with the idea that you might want to work with or learn from other groups or individuals. Although it is not a good idea to belong to more than one coven at a time, to be involved with more than one magickal group is not necessarily a problem. For example, I am the leader of a coven, and I also teach a Wicca 101 class for an online school. Several friends have asked me to lead a series of rituals with them, to help them explore their spiritual path outside of a coven dynamic and we meet every other month to do so. None of these events conflict with one another, and so I maintain a healthy balance.

▸ Does the coven have by-laws or other documents that you can examine before you join? If they won't show them to you until after you've been Dedicated or Initiated, be wary; these should be fairly public documents.

▸ Does the group ask you for more than a small amount of money to join, receive training, or be Initiated? Are gifts expected or required?

▶ Is the coven disorganized? Do meetings wander and typically end with a feeling of getting nowhere? Does participation and attendance at rituals not seem to matter?

You may have different reasons for being uncomfortable or wary of a group's true intentions. Listen to your intuition and follow your heart out the door, when necessary.

HOW TO FIND A MAGICKAL GROUP ONLINE

Traditional wisdom says that networking, or web-weaving, is your best tool in finding a coven. To find a physical coven, look in your local Yellow Pages for bookstores with metaphysical or occult books. The store may have a bulletin board of local pagan festivals, groups, or services. If you meet someone with similar interests, make contact and arrange future meetings. Ask him or her about the pagan community in your area and ask for help in finding a group. Keep your eye open for metaphysical shops, advertisements, and events in the local newspapers.

Online, you can join a list or newsgroup to make contacts. Find out if there are witches in your area whom you can contact and ask to visit their coven. Or check out the cybercovens that are now forming. Through my Web site at http://www.thevirtualpagan.org, you can contact online covens, make contacts within your community, and locate lists to join. Another great source is the network list on the Witches Voice Web site, located at http://www.witchvox.net/xwotw.html. Use this site to find other physical and cyber witches in your area.

Maintaining your safety and privacy is an issue, more so in the physical plane, and varying throughout the world. Be careful about how you make contacts. Do not give your telephone number or home address to just anyone you meet. Consider getting a post office box where you can receive mail under an alias or a Craft name. Get an e-mail account through Hotmail (http://www.hotmail.com), Yahoo (http://www.yahoo.com), or another free service and use it for all your Internet correspondence. Do not give anyone your personal e-mail address until after you feel safe and comfortable with that person or persons. If you are asking other people for assistance, insist that they honor your privacy and safety. If you are at all unsure about giving information to someone, then do not; again, trust your intuition.

CIRCLE ETIQUETTE—CYBER AND PHYSICAL

Each coven or circle has its own attendance policy, so the information that follows is general and based on my experiences in circle. Most of these comments will apply to both physical and cyber rituals.

/Preparing to Go to Ritual:

If your ritual information comes from an advertisement, then follow the instructions in the ad. Call the telephone number or send an e-mail to make inquiries and arrangements. If you have been invited to a circle that is not publicly advertised, please do not discuss it with others. Respect the privacy of those in the circle and never mention who was there to anyone else. You cannot assume it is okay to bring anyone else along, so ask the ritual or group's leader beforehand. Some groups are more open than others, but it is polite to ask first.

Do not attend a circle if you are under the influence of drugs or alcohol, and *never* bring drugs or alcohol onto the property of any person or group holding a ritual.

When accepting an invitation, always ask what time you should be there *and* what time the ritual begins. This way you can arrive in plenty of time to relax and center before the ritual. Ask what to do if you happen to arrive late. Barging into a ritual after it has already begun is a bad idea. Online, different groups have different attitudes about joining a circle late. When joining a cyber ritual, never announce yourself; it is the equivalent of shouting "Here I am!"

If you need to be somewhere else at a particular time after the ritual, then do not attend. I know that may sound unfair, but when we enter a circle we are between the worlds. Rituals take as long as they take, and that is almost always longer than you expect.

When you accept an invitation, ask if you should bring anything to the ritual (such as a robe, ritual objects, or an offering) and if you can help in any fashion. For a physical ritual, ask what you can or should bring for refreshments and if there is a fee to cover site rental, candles, or refreshments. For a cyber ritual, make sure you understand what items you may need to have on hand during the rite, and if there is a particular way your altar should be laid out.

If you bring homemade food for the feast, remember to label it with a list of ingredients, in consideration for those with special dietary requirements. If you are on a special diet, either bring your own food or talk in advance with the host or hostess to arrange for appropriate food. Some groups ask each attendee to bring a nonperishable item to donate to a local food pantry or other charity. Cyber rituals often include a feast time, and the leader may have a list of suggested items relevant to the celebration.

Ask about the dress code for the ritual. At large pagan gatherings, you are likely to see everything from elaborate costumes, to simple robes, to regular street clothing. At smaller public rituals, robes are preferred, but street clothes are acceptable. If you are invited to a circle that is not open to the public, then anything goes, so be sure to ask what to wear. If the ritual will be held outdoors, dress according to the season. Some groups may require participants to be skyclad, but robes are more common. In any case, if you wear street clothing, make it something loose and comfortable. Please do not wear t-shirts with pictures or messages that may distract others.

Certain types of jewelry may not be appropriate. For example, in some traditions, a silver Moon headband or an amber-and-jet necklace is worn only by the High Priestess. If you wear one of these to a large public ritual, you may be asked to perform some ritual duty for which you are not prepared. If you wear it to a private ritual, you could inadvertently offend someone. Obviously, these considerations are not important for a cyber ritual.

When in doubt, ask. Pagans tend to be nice people, and most of us remember what it was like to go to our first ritual.

/While Waiting for the Ritual to Begin:

If you have a request for healing or other magickal work, discuss it privately with the leader before the circle is created. Also, remember to visit the restroom before circle. Although you may leave the circle in an emergency, it can be distracting for others, and you may miss something!

In a physical ritual, do not touch another person's jewelry or objects without asking first. Many people go to great lengths to consecrate and charge their belongings, and this may have to be completely redone if someone else touches the object.

For all rituals, please remove watches, beepers, and any jewelry and/or ferrous metals that do not have a spiritual or medical purpose. Turn off your cell phone or beeper. Remove your shoes, if possible, as it is much easier to ground when you're in touch with Mother Earth.

In cyber rituals, there is a tendency to connect more directly to the energy flows through the chakra at the top of the head, which may cause you to feel dizzy or get a headache during ritual. Wearing a hat or covering your head with a scarf will moderate the energy flow if you feel it is too intense.

No smoking, eating, drinking, or chewing gum while in line or in circle. Smokers, please ask where smoking is allowed and never throw cigarette butts on the ground or into ritual fires! If you are participating in an outdoor ritual, apply insect repellant beforehand.

Any talk while in line for circle should be quiet and limited to questions regarding the ritual or warnings such as, "Don't trip over that root!" You can talk of mundane matters at some other time. Center and ground yourself while waiting in line. Use the slow walk to the ritual area to make a gradual shift of consciousness from the mundane world to the magickal realm.

Always shower before ritual. A handbath or bowl of water may be provided for you to rinse your hands in. You may be smudged with smoke, or you may be anointed with oil as you enter the circle. If you have allergies, feel free to let the anointer know, but quietly. These actions are for spiritual cleansing, which is an important step when entering sacred space. In a cyber ritual, you will most likely do this cleansing on your own, prior to logging on.

When entering the circle, you may be asked for your name, a password, or both. If you have a magickal name, use it; if not, do not be embarrassed about using your regular name, since plenty of people do. If you are asked for a password but have not been given one, you can try the standard "In perfect love and perfect trust." Online, your nickname is the name people will know you by, so you may want to change it when entering the ritual channel.

/During Ritual:

In physical rituals, you will most likely always move *deosil* (clockwise) when walking around the circle. In other words, when you enter the circle, you will turn left to move in the correct direction. When the Quarters (or four cardinal directions) are being called, turn and face that direction. If you are unfamiliar with the arm gestures and responses being used, just

stand quietly or try to follow along. In large public circles, you will probably see several different ways of saluting the Quarters, since each person
tends to use the method with which he or she feels most comfortable. One
or two pointed fingers can be a customary substitute for an athame or
wand in many circles.

Do not talk in circle unless the High Priest or High Priestess specifically invites you to speak. Stories, anecdotes, and discussion in circle are
strictly limited to Craft-related topics. Again, this is especially true in cyber
rituals, where every non-formulaic utterance is the equivalent of saying,
"Look at me." Out of kindness to other people, keep your story on topic and
as straightforward as possible. If you don't have anything to say, then a
simple affirmation of the mood or the theme is more than sufficient.
Remember that whatever you hear in circle is confidential. If you find
someone's story valuable and would like to share it outside the circle,
please ask permission of the storyteller first.

The High Priest or High Priestess may "Call Down" a god or goddess
during the ritual. This means that the deity has entered their body and
may speak through them. Do not talk at all during an invocation; it is one
of the most crucial moments in a ritual when mortals make contact with
the Divine.

In physical rituals, if you receive a small container of juice or wine, do
not drink it all immediately. Be prepared to offer a toast to the gods. It is
customary to offer part of your juice, wine, and cake or cookie to the gods
as a libation. A bowl or bowls may be provided for this purpose. If the circle is outdoors, the libations may be poured directly on the earth. Watch
what the others do.

If you feel faint or ill during physical ritual, please sit down or ask for
help; such feelings during ritual are not uncommon. You may also feel ill
during cyber rituals. If this happens to you, do not be embarrassed.
Remember that you are surrounded by loving and understanding people.
If you need to leave the circle during physical ritual for any reason, quietly
alert a member of the hosting coven and they will "cut" you out of the circle or explain their method for exiting and re-entering. Online, you may
simply imagine yourself pulling back from the web of the circle, rejoining
the loose ends on either side of you to one another. Then just leave the
channel. It is polite to send an e-mail to the ritual leader at a later point so
that he or she knows what happened to you.

/Post Ritual:

Rituals raise energy, hence the need to thoroughly ground the energy afterward. Online, the High Priestess or Priest will usually have the group place their hands upon the Earth, draining any extra energy down into Her. Eating your feast foods will assist in grounding, as will lovemaking and other vigorous physical movements. One note: I have found that the crown chakra remains energized for a variable period of time post-ritual. If another person touches this chakra, they may experience a nasty shock, followed by a sudden headache. Physically, this is a good time for hugs, relaxing, asking questions, singing, dancing, drumming, eating, socializing, and networking. Enjoy yourself, and make friends forever by offering to help clean up!

5: Communicating Effectively

When all of your communication takes place online in a text-only format, you quickly learn that there are limits to how understandable your posts can be. Humans converse with one another through physical cues as well as verbal ones, and much is lost when we are forced to communicate only through text.

STYLE BASICS

The online community compensates for the limitations of its text-based environment in creative ways. We use a variety of text styles to indicate our moods and attitudes. Unfortunately, most e-mail programs strip changes like bold and italic formats from text when they send and receive messages. Here are a few style basics:

- This is a normal tone of voice;
- THIS IS SHOUTING. Do not write your message all in capitals, as it is extremely annoying for your reader to imagine you shouting all the time;
- Sometimes, however, you can emphasize ONE particular word;
- You can use asterisks to place *more emphasis* on some text; alternatively, <use angle brackets>.

Acronyms are frequently used as e-mail shorthand. Common ones include:

btw	by the way
imho	in my humble opinion
imnsho	in my not-so-humble opinion
lol	laughing out loud
brb	be right back
kotc	kiss on the cheek
rotfl	rolling on the floor, laughing
rotflmao	rolling on the floor, laughing my ass off
rotflhms	rolling on the floor, laughing, holding my sides
wysiwyg	what you see is what you get
gigo	garbage in, garbage out
ttfn	ta ta for now
ttyl	talk to you later

The previous examples are some of the ways you can use text to keep communication clear. Other stylistic rules include:

▸ Do not keep the entire contents of a previous e-mail in your response unless necessary. Some people pay for their online time by the minute, especially outside the United States. Having to download long messages, which only repeat previous messages, costs more money.

▸ However, if you are replying to a message, especially a news posting, it is a good idea to include the relevant part of the original message. Including relevant information gives context to your responses. Quoted lines are generally marked with a ">" at the start of the line. For example:

>What is the name of that book with the whale and Ahab? I forget.

It's _Moby Dick_, by Herman Melville.

▸ Do not assume people know what you're responding to and who said it. In other words, if you are responding to an earlier post, make sure to include the relevant part of that post.

▸ If you want to talk about a subject, then start the discussion yourself. Do not harass others to begin talking about it. For example, instead of asking, "Why are not we talking about the upcoming Mabon ceremony?" say, "What shall we wear for the Mabon ritual?"

▸ Chain letters are frequently not acceptable and, in many places, illegal.

▸ Keep it short and to the point. Some e-mail systems do not relay long messages.

▸ Do *not* send junk mail. If you do not have anything to say, then don't waste bandwidth. Never e-mail material that the receiver did not request. You waste readers' time and money when you send junk mail, and it creates resentment.

▸ Be careful about sending attachments to a list, or any large group of people. If the attachment is large, downloading it could be expensive for some people; others may have e-mail problems with attachments. First ask who would like to receive the attachment, and then send it to only those members who respond.

▸ Cite your sources and use documentation. This practice lends credibility to your thoughts, as well as acknowledging those who thought of something before you did.

▸ Create single-subject messages whenever possible. If you have three separate things to tell your intended recipient, it may be better to send three different messages. Each of the messages can be filed, retrieved, and forwarded separately by the recipient (and sender), depending on the content. Subject lines in each message can show the contents of each message; such forethought allows quick scanning of messages in the inbox.

▸ Keep the list of recipients to a minimum. Consider an extreme but possible case: A message asking for comments on an article to be published contains a distribution list of twenty people. Each recipient responds and copies all original recipient, since, in many message systems, copying all recipients is the normal practice. If each recipient replies to each answer, 421 messages are generated, with the total system containing 16,421 messages. If each message takes an average of 100 characters, this process has used up 1.6 MB of disk storage, in addition, of course, to the human time and effort that has gone into this one e-mail.

COMMUNICATION STYLES

Although an in-depth discussion of personality types and communication styles is beyond the scope of this book, a brief overview is appropriate. Knowing who you are as a person goes a long way in helping to understand how others think and feel. Online, this knowledge is a vital component in your happiness with a group with which you have chosen to work.

When you try to characterize yourself or another person, you are stereotyping, which can be both useful and restrictive. I use two methods for initial descriptions, but I recognize that they are only valuable as an initial identification; I'm responsible for creating the deeper, more intimate communication that allows me to get to know each member of my magickal group.

The first method I use is astrology. The primary value of this system is that basic data (time, date, year, and place of birth) yields an enormous amount of information about the person and how he or she interacts with others.

The other method requires you to get to know the people whom you are going to type. I developed this after playing a game with my young cousin in which you try to guess what kind of animal someone is, and then combined it with my understanding of the magickal circle's quarters and elements. Each of the five points of the circle (east, south, west, north, and center) is associated with an animal. Most people fall into one of these definitions. Read through the descriptions and see if you can find yourself or others.

/Spirit/Center, the Spider:

In every group, some people always seem to be in the center of events, as the leader or due to circumstances. They are the spiders, the spiritual center of every circle. Spiders know what is going on, are close to most everyone in the circle, and frequently have the most up-to-date information about a situation. Spiders are not necessarily gregarious and outgoing; but everyone knows who they are, even if they don't know what the Spider does. Spider has a huge capacity for creation: connecting seemingly unrelated facts and forming new visions to share with the group. The Spider may be the leader, or just the person everyone calls for information. They are very good at looking for new alternatives to present impasses, as well as for seeing entangling situations.

Being a Spider is a heady feeling, sometimes, and Spiders can lose sight of the interconnectedness of the Web of the circle, placing their own importance above all others. They also have to avoid the self-delusion that being in the center of things can often bring. Having more information than others, Spiders may begin to feel like they know it all. The jealous hoarding of information can disrupt group communication.

Spider people can learn from all the others, if they open their many-faceted eyes to see the qualities of the other animals. Sometimes they lose sight of the larger picture, unless they pay attention to the Hawks. Overwhelmed by information, they may lose their enthusiasm for a project, unless reminded by the Jaguar. Whales help Spiders access their intuition, and Wolves provide the grounding and connection that may feel tenuous when the Spider hangs in mid-air all the time.

/Air/East, the Hawk:

The vision of the Hawk pierces through the fog of immediate situations and takes the long view. Hawk people are usually the ones who remind the group of its long-term goals and provide solutions for getting closer to achieving those goals. They may also have the best memory of what has already been done in similar situations. These birds are highly intuitive. Frequently, they can develop influence if it seems that they always have an answer for current problems. Their intuition can feel magickal to the rest of the group, as seemingly unrelated details are recombined to form clear messages and directions within the Hawk's astute mind.

Because they always seem to look ahead or behind, Hawks may feel flighty and a little out-of-touch to the rest of the group. Or they may blow in to meetings, solutions in hand, and feel rejected when the rest of the group takes a little time to understand why those solutions are the best. Believing that they know more than others in the group, Hawks sometimes refuse to see evidence that is staring them right in the face, so they sometimes make incorrect decisions.

Hawks can learn much from their circle companions. The Jaguar balances out Hawk's dispassionate observation with playful enthusiasm. The Wolf reminds Hawk to reconnect with the group at regular intervals. But from the Whale, Hawk learns about dealing with the emotional currents within a group. Hawks are least able to understand emotion, and they prefer to ignore its influences.

/Fire/South, the Jaguar:

Jaguars are sensitive to group energy flows, and frequently help people feel good, welcomed, and comfortable. Jaguars are almost always well liked and their enthusiasm can carry projects forward to completion, despite the fact that they tend to be loners, even within a group. If the group seems to be fragmenting, Jaguars can usually spot it first, and do the most to reintegrate people back into the circle. These people make excellent leaders in the early stages of projects; but are not always so good in the long term.

This energy can be too strong for some people and over-enthusiasm can keep a doomed project going for longer than is healthy. Jaguars need followers to maintain their energy and a negative dynamic can ensue when they refuse to let their supporters stand on their own. Enjoying energetic exchanges, Jaguars can seem to agitate situations, while remaining completely unaware of the disruption they have caused.

Jaguars need to be balanced by the dispassionate and far-seeing vision of Hawks, as well as by the emotional sensitivity of Whale. Wolf's grounding nature can teach Jaguars the most, since it reminds them that they are not the only ones with valid reasons and enthusiasm for a project. Being a leader requires the agreement of the group, not the tyranny of an individual.

/Water/West, the Whale:

Whale people excel at understanding and recognizing emotional currents within groups. They frequently become aware of conflict before anyone else. They uncover the sources of conflict, bring them into the open, and then mediate until the problem is resolved. By asking open-ended questions, Whales increase communication among members who may normally avoid one another; they increase the group's sense of harmony and oneness.

Like Cassandra from Greek mythology, Whales' ability to foresee conflict can make them unpopular with the rest of the group. If the Whale remembers that it is more effective to ask if there is tension and how others are feeling, then other members feel invited to speak, rather than being told how they feel.

Whales learn a great deal by occasionally acting like the Jaguar: enthusiastic about a project, rather than immediately pointing out the potential problems. The Hawk's overview provides the most balance and wisdom for

Whales, who are frequently so caught up in the intimate emotional moment that they lose sight of the big picture.

/Earth/North, the Wolf:

Wolves are at their best when providing for the group's needs. The most practical and realistic of the signs, they guard resources and protect the group's boundaries. They tend to see the group as their pack and can be sensitive to changes in leadership. Wolves remind us to remain individuals, with unique value and importance, while supporting the group as a whole. They are frequently the ones who encode the knowledge gathered by the group, seeing it as a valuable and unique resource that must be preserved and protected.

Wolves are least valuable when they act like their domesticated children, dogs. They sometimes follow the rest of the pack and ignore individual needs and desires. If they forget the unique value of each member of the group, Wolves may support whoever is in a position of knowledge at the moment. They may loose sight of the shared vision and focus only on the one presented by an energetic individual.

Wolves must learn to accept that emotions and vision are as valuable to the group's needs as any material resource. Shared knowledge is vital, but must be added to prevent stagnation. The Jaguar's enthusiasm and self-reliance are traits that benefit the Wolf—especially if the Wolves forget to assert their own viewpoint as valuable and subsume it into the larger group vision.

I realize that the animals I work with in this model may not fit your personal ideas. Feel free to explore your own understanding of how the people around you fit (or don't) into animal roles. Remember that just because a person seems to be a perfect Jaguar, does not mean that he or she will be that way all the time. People change over time.

SPAM

Spam is a nasty side effect of the open community of the Internet. Succinctly, spam is what happens when someone sends an unwanted message (usually commercial or political in nature) to a large group of people. There is, however, another form of spam that is more difficult to discern. One of the complaints from pagans who have just joined the pagan community on the Internet is that, after doing something that they felt was helpful, they get nasty e-mails (also known as "flames") from some of their brethren. What they have usually done is perpetuate one of the many Internet hoaxes floating around.

I know of no other way to put this, so I will be straightforward:

▶ Any virus that threatens your operating system or browser is usually posted on the homepage of your operating system or that of your browser manufacturer;

▶ There is no Internet tax about to be passed;

▶ The spam about NPR funding is about five years out of date;

▶ The one about Proctor & Gamble being a tool of Satan is about ten years out of date;

▶ No company is offering to give anyone $1 for every e-mail it receives;

▶ Furthermore, no child has ever made a wish to get a million e-mails or postcards;

If you get an e-mail that says, "Urgent! Pass this on to as many people as possible," delete it. Rarely do these e-mails have any value. If you are in doubt, five minutes of research can clarify its truth. Just take the title of the e-mail (without all the FWDs and REs), and paste it into a search engine. If that doesn't work, find the Web site of the group that the spam supposedly benefits and see if the information is readily available. For instance, the American Cancer Society has a great page denying the "little girl dying of cancer" spam. Two excellent sources are the National Fraud Information Center (http://www.fraud.org) and Hoaxbusters (http://HoaxBusters.ciac.org).

As a warning about privacy, when you receive an e-mail as part of a group of people, if a single person on that list sends that spam on, your e-mail address is included on that forward. This is one of the ways that computer-disabling viruses are spread throughout the Internet. If you receive such a message and do not wish to have your e-mail address passed on, you must reply to everyone who was sent the e-mail. Be polite, and realize that some of them are going to take offense. Ask that, if they choose to forward the e-mail, please delete your e-mail address first. You might also remind them that, if they choose to send it to multiple people, then simply using "blind cc" will keep others from having to send out such messages.

Here are a few warning signs that a message is spam:

▶ "Forward to as many people as you can" appears in the text of the message;

▶ The message is from someone you do not know;

▶ The message is from someone to whom you cannot reply;

▶ The message sounds exactly like a known scam: little girl dying of cancer, lung cancer, or heart disease;

▶ The message uses abusive language: "Only a real jerk would delete an e-mail that was this important!";

▶ If sent to a mailing list, the message is utterly unrelated to the topic, or to mailing lists in general.

Use your common sense to deal with spam.

BODY LANGUAGE AND EMOTION

The online community tends to be very literal about using text labels to indicate emotions or facial expressions. This protocol has developed for two reasons:

1. People feel strongly about a subject and want to express the strength of their feelings.

2. Many examples abound of messages in which emotions were misinterpreted or confused with the other content of the message.

If you label attempts at humor, anger, or sarcasm as such, those feelings can be transmitted with less chance of misinterpretation. For example, saying "You are such a pig!" can be read as insulting, or (with a small addition) humorous: "You are such a pig! :-)" Another example would be the use of single words in between brackets to indicate emotion and body language, like [sigh], {smirk}, or <wink>.

When you're talking to someone online, it can be hard to interpret the ideas behind the words, the writer's feelings, and the context. So, to get past this problem, emoticons have emerged. Emoticons are groupings of symbols that, when viewed sideways, form different expressions (with a little imagination). Most of us remember the "smiley" rage that took over

popular culture not long ago. In some cases, people use emoticons to indicate personal attributes.

The general construction of a smiley is easy. First you start with the
eyes, these can be one of a group of things, including the more commonly
used ":," "8," and "=." Some people add a nose, for instance "-" or "*." Then
you simply add a mouth to finish the smiley and you can communicate
more than words alone. What follows is a list of some of the more commonly used emoticons:

:-)	smiling
;-)	winking
:-(	frowning
:-o	astonished
8-)	wears glasses
:->	a little wicked
%:->	curly hair
5:-)	Elvis
=|:^>	wearing a top hat
<g>	grin
<BWG>	big wide grin
<BWEG>	big wide evil grin

Countless other potential combinations exist and you can always make up
your own. One of my favorite lists of emoticons is The Unofficial Smilie
Dictionary at http://www.charm.net/~kmarsh/smiley.html.

6: Connecting with One Another

The previous chapter explained some of the ways in which online interaction is handled textually. However, the dynamics of online group interactions are more complicated than those found in text. This chapter deals with these issues. For some people in the world, the tangled intricacy of people in groups is as clear as a summer morning; I am not one of those people. My education evolved through more than two decades of pagan activities, as well as my mundane interactions in different work places. I have learned mostly through trial and error how group dynamics work.

FORMING THE GROUP MIND

The most common question about how cybergroups work is: "How connected do members feel, compared with the sense of unity in a physical coven?" The answer is that the sense of connection depends on the interaction among the coven members. Another way to ask this question is, "How do you form a strong bond when there is no physical contact?"

Group magick requires that the individual psyches within the coven meld, forming what is called the "group mind." In *Applied Magic*, Dion Fortune says, "[T]he Group mind is built up out of the many contributions of many individualized consciousnesses concentrating on the same idea."[1]

Once formed, this group mind has its own momentum and can amplify the power of individuals or a group. This consciousness has no

awareness of its own being. The group mind focuses and directs the combined energy of the coven. As a result, the power and efficacy of magickal
actions increases exponentially.

Physical covens generally meet monthly, weekly, or sometimes only to
celebrate the Sabbats. In between, members socialize to increase their
bonding and connections with one another. In some groups, this fellow-
ship is enough to create a sense of unity so that magickal workings are
clear, focused, and direct. Sometimes groups simply don't gel, in which
case, magickal workings are not as powerful. Covens with a greater sense
of unity most likely spend time at every meeting strengthening their identity and doing meditation as a group.

Cybercovens meet almost daily. Perhaps not at the same time, and not
always with something to say, but we are sharing energy with each other,
imparting information, and bonding. Cybercovens use a slower and also
more intimate method of communication. I have no idea what
GorgonsDaughter looks like, but I know about her children's troubles in
school and her personal rituals that she performs to modify those troubles.
I know what her sacred space and her altar look like because she has
described these to me in detail. Each of us takes turns, for a week at a time,
posting a file to read and discuss. This file may be a piece of text, an image,
or even music. These meditations focus our group attention on a single
idea, creating a constant flow of energy moving around the world, never
ceasing, and increasing in strength with each new day.

This flow of energy is magick in its purest form. For example, I post a
message in the evening asking coven members to light a candle and send
energy to my sister for a few days because her bar exam is coming up.
Members online with me in the West read the message and the energy
begins to move. Paladin, in Australia, is the next to read it (technically the
day before!) and she adds to the current, then Brightayes in Germany, followed by CougarStar in the UK, and so on, until the energy has finished
its journey around the globe, increasing in strength and intensity.

These energies bond us together within the circle. We become a group
of networked individuals, like a grove of trees in the middle of a field. To
the outsider, each tree is a separate organism; but, underneath the Earth,
our roots merge and we become one organism. We support one another
and grow stronger; the group mind grows stronger.

SHARING

Whenever a cybercoven accepts new members, it is a good idea if everyone shares stories about who they are and why they joined the coven. The telling of these stories begins to weave the members together into an inclusive tapestry. As part of this sharing, the High Priest or Priestess might assign exercises to expand personal horizons and help coven members see ordinary objects in new ways. For example, you are told to sit in a darkened room with a candle burning before you. Sketch what you see while in a meditative state, then describe the experience. What did you learn? What you drew does not matter, your perceptions are altered by the meditative experience and you become more open to the energy of the unseen world.

Another excellent exercise is to have the teachers share daily writings or graphics with coven members. Each piece provokes a response or comment, thereby creating a pattern of daily thought that every member shares. In other words, each day the coven members are thinking about the same things, further binding them together.

Sometimes, the exercises are designed to bring us together, to change our group dynamics so that we are more unified, more constant in our energy. When a coven member tells the others why he or she is here and what he or she wants from this experience, the group mind is strengthened.

EXERCISES

A cyber group can use a variety of exercises when it accepts new members. Generally, the focus would be on having fun and getting to know one another, although some exercises are planned to help the leader gauge each member's level of magickal knowledge and perceptions. One exercise is to hold daily question-and-answer sessions. Here are some topics to explore:

▸ What is magick? When have you done magick? How did it make you feel?

▸ What is the single most important benefit of magick?

▸ What do you think it means to be a witch?

▸ What does the ability to use magick mean to you personally?

▸ What is the God or Goddess to you? How do you see Him, Her, or It?

▸ What family and personal rituals do you perform?

▸ Describe your most powerful spiritual experience.

▸ Think about what you would like to change in your life and explain how you would use magick to make these changes.

▸ What experiences have you had with extrasensory perception (ESP)? Ghosts? Psycho-kinetics (PK)?

These exercises and others like it allow members to become more than just text on a screen; answers flesh us out, make us more real. I remember the time I spoke to Traveller on the telephone one evening—it was an absolute shock, because, although I knew that she was from Texas, I had never heard her accent. She told me later that she was surprised to hear me giggle. My online personality is a bit more formal than my everyday persona.

A cybercoven's daily communication enhances each member's ability to integrate magick in our lives. Most of us live such splintered lives. To the outside world, we are parents, teachers, or employees, but in the private world, we are witches. In a physical coven, we reaffirm our beliefs through ritual and regular meetings. If life intervenes because the baby is sick, or work makes its demands, we miss that time completely.

To further a cybercoven's cohesion, I recommend performing ritual frequently. "What could be more conducive," asks Fortune, "to the formation of a powerful group mind than the secrecy, the special costume, the processions, and chantings of an occult ritual?"[2]

Cybercovens have their unique advantages. Life can still prevent us from attending a ritual or meeting. The sense of continuity is very difficult to achieve and maintain at a steady level in a physical coven; but a cybercoven offers the chance to interact in a different dimension.

No longer bound by the physical laws of distance, merely the fourth dimension's law of time, cybercovens include far-flung members who add a richness and depth that we would otherwise never share. In a physical coven, I would never have known about John Tallberries' mountain in Tennessee, because I live in Seattle. Nor would I be able to share CougarStar's exquisite British Traditional knowledge, because she lives in

England. Through joining a cybercoven, we are all members of a community, bound by our belief and our faith, not by where we happen to live. Our broad background of life experiences becomes clear as our conversations include topics like social policies, ethical considerations, economic decisions, and our perceptions of the Deity.

FLAME WARS AND OTHER MISCOMMUNICATIONS

This section might be subtitled: "Lady, Why Is My Robe on Fire?" I have spent time in a variety of groups and circles, and have spoken with many pagans over the years about their experiences. A few consistent patterns in the life of a coven have become clear to me:

▸ Most covens do not last much longer than five to seven years;

▸ Covens have a constant, though varying, rate of turnover;

▸ More people join covens than stay;

▸ Covens do not break up because of theology.

The cybercoven community is young, not even entering adolescence now. Those of us who have been here for several years have seen the truth of some of the previous patterns. Only time will tell about the first pattern mentioned.

Like most groups, covens have problems that are the direct result of miscommunication, misinterpretation, anger, strong-willed personalities, and power trips. In fact, the most mediocre coven can produce plots that could be the basis for a six-figure-a-year soap opera script.

The truth is, humans are basically social creatures; we like being with people. Unfortunately, our expressions are not perfect and we frequently say one thing and mean another, or hear something and think we understand, only to find out later that we were mistaken. Some of us have nurtured our intuitive abilities to such a degree that we read body language to inform the knowledge we gain from other methods. In the physical world, using multiple forms of communication increases our ability to get our message across, as well as to understand what we hear.

In the cyber realm, communication is one-dimensional: the written word. We have a hard enough time understanding people face-to-face; that difficulty is multiplied a thousand-fold when we read e-mail.

Which brings us to the flame. Anyone who has spent even a small amount of time online knows about the flame. You send a post, maybe in response to something you read. You just dash off your opinion on the use of chlorine as a household cleaner, for instance. Next thing you know, someone is reading you the riot act because you promoted a deadly agent as a normal part of life. She calls you unethical, a killer no less! Well, you are not going to sit still for that, especially since just last month she was telling everyone how well smoke works at disturbing the bees she keeps. In fact, that proves she is no pagan, since she is willing to disrupt the natural order of things. And so on.

Feeling warm?

Part of being a pagan is that we accept responsibility for our own actions, including our communication. When we speak with courtesy and respect, we act out of honesty and trust. Our connection to one another is enhanced and the group flourishes. We have the responsibility to communicate our frustration with another coven member directly and avoid getting caught up in he said... she said... I heard. Criticism delivered in the form of, "Other people are upset with you, but I cannot tell you who, just why," is upsetting, frustrating, and unfair. A more positive approach is to role-play a conversation or review the message before sending, giving positive feedback, but not getting directly involved.

Because 99 percent of our online communication is through the written word, problems result. For example, the absence of auditory or psychic nuances. Comments such as, "You are an idiot" said affectionately with a smile are received quite differently from "YOU ARE AN IDIOT!!" with no other cues. Writing is a skill at which not everyone excels. Sometimes readers interpret e-mails as vague or contradictory when the content seems crystal clear to the author. Or a comment that seems commonplace to a writer can push all the wrong buttons for an audience. Anything written only captures a moment in time and we all have had moments we later regret.

One characteristic of cybercovens is that they are much easier to leave than physical covens. If Cindy the Chlorinator is feeling attacked, she can enter a few commands and we are all out of her life forever; she will never know about more moderate replies or invitations to return. In a physical coven, if a member disappears, another member may call or stop by the house to make sure he or she is all right. Online, it is easy to eliminate everyone except those who agree with us in everything. Learning from each other is difficult. If we truly believe we have nothing to learn from those who have different opinions on some issues, we can be certain that most of our opinions are not based in fact.

GUIDELINES FOR POSITIVE COMMUNICATION

Christine Baldwin, author of *Calling the Circle*, describes the circle as a place where people accomplish successful problem-solving, give and receive nurturing, and appreciate or are appreciated from the heart. Further, she describes the circle as a place in which differences arise and are worked out, supported by conscious awareness and encouragement by all. A strong coven is one in which this type of dynamic and loving inter-action takes place, particularly when difficult issues arise.

A valuable resource for creating mutual agreements and understand-ing is the Compact or Agreement. This document defines how the group agrees to behave, self-govern, and take personal responsibility. JaguarMoon's Compact begins with a statement of purpose, briefly describes the tradition that the coven follows, and then lists the five oper-ating rules with which members agree: respect, honesty, confidentiality, accepting responsibility, and attendance. It is reviewed annually so that it may reflect the changing needs of the coven. As a separate document, The Art of Ritual Class Agreement lists twelve things expected of members, including the promise to take care of themselves and their family before class matters. Both documents can be found in appendix B on page 140.

By agreeing to the same ground rules, members start from the same place and use the same tools to communicate with one another. These agreements promote a feeling of safety for each member. Members know what to expect and what is expected of them. But these are the common-places of group work, and are not sufficient. These five specific steps ensure smoother, cooler, communication online.

1. Make I statements versus you statements: Speaking for yourself and saying how you feel is always better than telling someone else what you think of him or her. When you attack others, they naturally go on the defensive. Talking to someone through a wall is much harder than talking face to face. For example, "I hate it when you barge right into discussions and take over. It makes me feel as if I should not even bother," will be received differently than, "You are so rude! You come rushing in here and take over as if nothing else is going on!" Beginning a statement with "I feel" gives information and lets other people take responsibility for their actions while you take responsibility for your own feeling.

2. *Avoid absolutes:* Couching your statements in extremes just makes what you say sound absurd and dismissable. "You always barge in and never let me talk!" versus "I feel that you frequently interrupt me and take over the conversation." The first is giving a reason to eliminate the person from your life. The second is an invitation to solve a problem.

*3. **Think before you post:*** I know it sounds basic, but before you send off that blistering post to Nyghtwynnd about her complete inability to recognize the danger of chlorine, stop for a moment. Read through what you have written. If you were reading this, would you be offended? Perhaps this post is best sent privately and not to the entire group. If you are angry at all, then do not send the post. Wait a day. Talk it over with a loved one, or your mentor, or another coven member. Check your signals: You may have misinterpreted or misread something. Wait another day. If you are still angry, read your post again and maybe make some changes. Always wait to send a post if you are offended and hurt. Ask yourself, "What do I want to accomplish by this post? What am I likely to accomplish by this post?" If your first draft of an angry response says exactly what you want to say, you are probably still too angry. You lose nothing by waiting. It is almost never necessary to have a public disagreement. Disagreeing in public is an attempt to:

- ▸ Marshal support for your side;
- ▸ Get rid of your opposition;
- ▸ Force the group to decide which of you should leave;
- ▸ Split or destroy the group.

You are responsible for your communication. Being right sometimes entails being hurtful, so, be careful.

*4. **Remember the human factor**:* The timeless wisdom of the Golden Rule states: Do unto others, as you would have others do unto you. Stand up for yourself, but try not to hurt people's feelings. It's ironic that computers bring people together who'd otherwise never meet, but online interaction changes that meeting to something less personal. Humans exchanging email often behave like some people behind the wheel of a car: they curse at other drivers, make obscene gestures, and behave rudely. Most of these people would never act that way at work or at home. The interposition of a computer seems to make rudeness acceptable.

Yes, express yourself freely, explore strange new worlds, and boldly go where you've never gone before. But remember that real people are on the receiving end of your communications. Would you say what you wrote to the person's face? Of course, it's possible that you'd feel great about saying something extremely discourteous to the person's face. In that case, reading this chapter will not help you. Go get a copy of *Miss Manners' Guide to Excruciatingly Correct Behavior.*

5. One last rule: From *The Notebooks of Lazarus Long* by Robert Heinlein: "Never argue with an idiot. Other people may not be able to tell the difference."[3] Another way of saying this is, if you argue in public you never enhance the group's respect for you, and you will probably make an ass of yourself.

Remember: Anger is a bad way to teach someone a lesson. As Heinlein said, "Never try to have the last word. You might get it."[4]

7 : Potential Problems

I have spent the last six chapters talking about connecting and communicating with other pagans online, but now it is time to discuss the uncomfortable things. If you thought that participating in a cybercoven would free you from backbiting politics, unfulfilled expectations, or the lack of cohesiveness that can plague groups, you were wrong. Hopefully, my experience gives you some idea of what to expect, what to do, and how to avoid problems.

The three categories of problems are: cohesiveness, politics, and continuity.

COHESIVENESS: MAINTAINING THE GROUP MIND

Cohesion is the sense of belonging, of participation. Some groups quickly achieve a cohesive identity; others struggle to find it or ultimately disband. Although creating a sense of cohesion is a problem for physical and cybercovens, I believe it is more difficult to achieve online. "Perfect Love and Perfect Trust" is a foundation element within any magickal circle, but it cannot be assumed from the start. I never immediately trust and love a person I have just met, no matter how good his or her energy feels, how many past lives I am certain we spent together, or how much he or she likes me. Contrast the positive vibrations of my mythical encounter with the cool impersonality of the text-based online environment and you can see why achieving cohesiveness is so difficult for cybercovens.

However, if you look at "perfect love, perfect trust" as an ideal, and separate it from the concept of intimacy, it is easier to understand. Intimacy is not sexual; instead, it is accepting another's nature, while at the same time being aware of your needs, and stating them. Intimacy is closer to equality than sexuality. True intimacy cannot exist in the absence of trust in self or others. If intimacy is present within your group, then your circle, sacred space, and your love and trust allow you to practice magick and connect with the Divine. It becomes an expression of our achieving union with our Higher Self, that part of us that acts in accord with the highest purpose of the universe, as well as with the God or Goddess we love and worship. In a sense, entering the sacred circle allows a witch to shed all negative personal aspects, and become closer to "perfect." This process does not work if you are in discord with another member of the group, but I will talk about that later.

Witches do magick, a mysterious process that begins in the unseen world and manifests on the physical plane. For a group to do magick, each person must be focused on a specific outcome. To be focused, in part, requires that you trust that other coven members act in accord with the outcome.

You cannot force intimacy; it grows gradually and requires nurturing over time. If you try to prove perfect love and trust, you invite the dissolution of your group. As an example, in our first year, ShadowMoon was given a long list of laws by which we had to abide. We were given the opportunity to read them all and to choose some that we felt were inappropriate. A few members took that opportunity, but most did not. Then our High Priestess revealed that the laws were false, deliberately designed to make us speak up and out against them, that this lesson had actually been one of "Always question." Although most members understood the lesson, a few did not and they left the coven after exchanging angry words and nasty insults. It was an unpleasant experience. We have never repeated that lesson; the price was too high.

Intimacy evolves through information sharing, recognition of the other's individuality, the exploration of one's self, and mutual respect. You cannot force this process; rather, you must let go and allow it to happen of its own accord. It takes a minimum of three months of continual exercises and participation online before cybercoven members feel a true intimacy and a sense of community and cohesion. Online participation is often on a daily basis, unlike physical covens that only meet periodically.

Some members may participate in the process of intimacy and yet never feel as if they are members of the same group. I have had people tell me they were leaving, after spending their year and a day in ShadowMoon. When I asked why, they said that they never felt that they had formed deep bonds with the rest of the coven; they still felt like outsiders. Although painful, it was a valuable lesson for me to observe that each participant perceived the same experience so differently.

Online, cohesiveness needs constant maintenance. This cohesiveness is a vital part of a cybercoven's continued existence. Real life always interferes with online participation and, unless its members actively work to maintain their bonds, the coven dies. All coven members have personal responsibility in creating and maintaining cohesiveness. If you are a naturally withdrawn person who prefers to have others ask if there are problems, then you have to work much harder to maintain a sense of community. You cannot see a person withdraw online, nor can anyone see that you need someone to reach out to you. You have to do some reaching yourself. Some ideas for encouraging cohesiveness include:

1. Do ritual. Doing ritual, no matter how small or simple, is a powerful way to build community. Each time you work together magickally with other coven members, you add to the group mind and the dynamics of the coven's structure.

2. Encourage sharing and storytelling. The best way to do this is to share stories from your own life, or things that influence you. You can also do:

▸ Daily meditations. Each week, have a new person take his or her turn at posting text or graphics that are noteworthy. There are no rules for content other than the general ones of politeness. Content varies and can include quotes from the world's philosophers, heavy-metal song lyrics, and passages from books and essays on how to organize your life.

▸ Photo meditations. As with daily meditations, the author changes each week and there are no rules as to content. These meditations have ranged from a hilarious series of "anti-motivation" posters through computer artwork, to the paintings of Masters and personal photographs.

3. Pen Pals. Have a coven elder randomly assign each coven member to another for a month. Encourage them to write to one another regularly,

share stories, and just become acquainted. Frequency and depth of participation are up to them. If one is not happy, both should try to work it out
before going to the elder. This system of sharing should encourage personal bonding within the coven.

4. Create a buddy system. Divide a large coven (one with more than thirteen members) into smaller groups within the coven, headed by the
coven's elders, and include the High Priest or Priestess in each group. Each
buddy pays attention to the others in his or her group. If one of them is
not posting, or is only posting without content, the buddy contacts the
newcomer to ask if everything is okay. This system allows shy members to
be drawn out a bit more. Moreover, if a member is feeling depressed or
alone, a friendly message asking, "Are you OK?" can help. If yours is a
teaching coven, you probably will not have to do this, since each student
will have a mentor.

POLITICS AND CONFLICT:
THE UGLY SIDE OF GROUP DYNAMICS

The ugly side of politics has probably existed since the first meeting of
more than three people. You would probably expect fewer issues in a spir-
itually focused group; unfortunately, that is not the case. The truth is, magick attracts a lot of people, and some of them are only interested in building up their own personal power and supporting their self-image. These
people are the troublemakers in every coven, and they are not always easy
to spot.

As an example, in its first year, ShadowMoon had an issue with a well-
known member of the online community. She was a self-proclaimed third-
degree witch, and acted as the administrator for the e-mail list that was the
genesis of ShadowMoon coven. Our High Priestess asked her to join the
coven and to act as her HandMaiden, and this witch enthusiastically
accepted. As the months went by, however, she did not post lessons, rarely
attended rituals, and did not participate in discussions. She was essentially
a nonmember. Finally, the High Priestess spoke to her privately, pointing
out that she had duties to the coven that were not being carried out. This
witch arranged to have that private call secretly monitored by a third party.
The witch guided the call so that confidential information was revealed,
which was then publicized to the list. A situation ensued in which a lot of
people suddenly doubted the validity of what they had been taught and the

trust of our coven was shattered. We had to close down the list and move to private e-mail to continue our correspondence. Several members left and the outcome was devastating to the coven's energy.

The lesson? On one hand, the HandMaiden just wanted to be able to say, "I'm a member of an online coven," without actually participating. This experience was also a lesson in trust. No matter how well you like people, get to know them first and watch how closely their actions match their words. Then you can judge whether they are worthy of your trust.

Another situation arose when a coven elder encountered difficulties with a physical coven in which she and her husband were members. This coven practiced a tradition in which sex with people other than her partner was a required part of ritual and the leaders held control through mental manipulation and tactics designed to destroy members' self-esteem. When our group heard of these practices, we were horrified and offered her love and support while encouraging her to leave the other group. Instead, she saw our actions as interference, accused us of being nosy (although she had shared the information with us of her own accord), and launched a nasty e-mail campaign against the coven leadership. We had to ask her to leave. Her pain and self-hatred was being turned against us and was destructive to the coven.

The lesson? Some people can never let go of their own pain; it is too reassuring for them, too comforting. To release it for self-love is too difficult and you are powerless to help.

When someone leaves a group, the people remaining may feel shaken, abandoned, or threatened. They may, of course, also feel relieved. If many people abandon a group at once, the group may not survive. When one person leaves a group, it leaves behind an implicit statement that the group or its members are not valued. Of course, the person who leaves may not mean to make this statement: he or she may have valid reasons for going. Formal leave-taking or a ritual can help the group to recover its integrity. Unfortunately, many people leave groups by simply drifting away, without ever voicing their anger, criticism, or appreciation. If you do value a group, please give the members a chance to say good-bye.

Personally, I experienced a different kind of political situation that arose out of a misunderstanding that I had with another elder. I am a natural organizer and, over time, I began to take on a lot of coven duties related to organization. One member felt threatened by my assumption of what she perceived as her duties and, therefore, her authority. This situation was exacerbated by the fact that my tone online tends to be a bit direct. Other people have called it pedantic and authoritarian. Apparently, my tone irritated her. She began to snipe at me on the list, pointing out small errors, and treating me as if I were incompetent. She was using the list as a forum for her discontent. The situation was becoming destructive, involved other people, and some were being forced to take sides. Luckily, we had an epiphany about each other and our viewpoints, and we were able to reconcile our differences. But it was luck and a touch of divine intervention that did it; otherwise, we could never have become friends.

The lesson? Text-based communication is easily misunderstood and the responsibility of clarity is on the sender and the reader. I can only speak my truth; if you do not agree, you must tell me so I can find a different way to explain. Or we must agree to disagree.

Two kinds of conflict occur: those in which people come together to resolve the situation, and those in which people escalate the animosity. Frequently, the issue over which the conflict arises is far less important than how we solve the problem.

Part of the reason that all of these problems arose was the lack of physical contact. I am more warm and friendly in person than I am online, and if the elder had known that, she would have realized I was just trying to help rather than undermine her authority. If we were all present in the same room, we could have better understood each other's true intent through tone of voice and body language. By asking, "What's the matter?" we would have immediately detected a problem instead of "reading between the lines," fuming in silence, and getting ourselves and others wrapped up in an overblown issue. So in a cybercoven, it's extremely important to be clear and to ask questions when you sense negativity.

Some warning signs of a coven drowning in negative politics are: apathy, lack of participation, low attendance at rituals, promises broken, gossip, backstabbing, and flames. If you find yourself in a coven where any of these factors exist, walk away. You cannot fix it. Remember that leaving in anger does not help the coven or you; so just leave. If you absolutely must say something, just be brief and to the point, and remember that your words may come back to haunt you.

Covens must be aware of the use and misuse of power when only a small group controls the coven—not just in cases where a leader controls everything, but in subtle ways as well. For example, you have created a cybercoven and your best friend is a witch with decades of experience. You ask her and she agrees to join the coven and share her knowledge. She immediately begins to hold classes and do rituals and you're left with nothing do. A few months later, some of the coven members write to you privately about how bossy she's gotten and ask, "Whose coven is it anyway?" In this case, an honest discussion between the two of you will probably clear up the issue for you both, but maybe not for the rest of the coven. Because, even if you deliberately set out to equalize the power, it requires ongoing awareness and attention to keep it balanced.

A trusting, safe environment, where intimacy is nurtured but not forced and sharing is spontaneous and loving, produces a coven of extraordinary strength and power. A coven is frequently likened to a family. There is a mother and father (the High Priestess and Priest), older relations (the elders), and younger siblings (the rest of the coven). Squabbles happen, but they are dealt with, and misunderstandings are quickly solved and forgotten. The coven becomes the place where members turn for love and support. The deepest trust is built out of the conflicts we resolve, for how can we know the strength of the structure we build until it is tested in strife?

A coven is also a group, and subject to all the issues that arise when people from different backgrounds, holding different points of view, get together. This diversity is much harder to see than ugly politics. In the natural process of a group, members move from being warm and fuzzy (where people are on their best behavior), through chaos and confrontation (as everyone gets real), into a sense of true community. Living through that middle stage is hard; it requires an effort to be present, loving, and honest.

People have prejudices and, although the cyber realm offers no visual cues about one another, it does not erase our mental conceptions. For example, one coven member felt that I had a "high and mighty" personality. Her comment reflected the fact that I remind people to get things done and not fall behind. For a time, I saw her as a bit of a slacker who put her coven duties last. Over time, we realized that we were unhappy with how we saw one another and made an effort to get to know each other better. In the process, we learned to communicate better. Now when I send her a reminder, she is grateful.

Then there are my favorite difficult people who are passive-aggressive in their behavior. They are the people who are unhappy, but never say anything directly. One of our founding coven members had been gradually withdrawing over a period of a year until her participation was nearly nonexistent. I sent her several direct messages to ask what was going on. Finally, the High Priestess stepped in and sent her a message telling her to call collect, if necessary, but to contact her immediately. She did not and so, a day later, we removed her from the coven. She sent us an angry, vicious message bemoaning the coven's decline into "just a list" where "no one has any connection to anyone else," unlike the early days when we were a "real" community. We all thought this was misguided behavior. We wondered how she could feel connected to a community in which she did not partic-ipate. No one could understand why she stayed when she was so unhappy.

I am not going to leave this section on such a sour note. Ways exist to deal with nasty politics and conflict. These guidelines do work:

1. Use "I" statements when talking about your concerns. "I" takes respon-sibility for your feelings, rather than accusing or judging the other per-son.

2. Speak from your heart, say how you feel. Feelings are valuable and shar-ing them is a powerful act.

3. Be clear about how you feel before you speak. If possible, separate your feelings about the person or the manner in which the issue was pre-sented from how you actually feel about the issue.

4. Deal with the issue in the most appropriate place. If someone says something rude to you on the list, do not respond in public; it's the equivalent of getting into a fistfight in plain view. Take it to private e-mail. If you are concerned that your words will be taken out of context, copy someone you trust on the message.

5. Deal with the issue as soon as possible. If you are gathered in a meeting and a proposal is made to do something you think is a bad idea, do not wait to speak with people privately; raise the issue immediately. Do not let time pass before you contact the source; your anger will grow and fester. As well, the more time that passes, the more likely it will be that the originator has forgotten the context of the situation or what was said.

In the end, remember that what makes conflict a valuable lesson is the will-ingness of everyone involved to resolve the issue and learn from the expe-rience.

CONTINUITY: KEEPING THE CIRCLE VITAL

Of all the problems in an online coven, continuity is the most difficult. As the coven moves through stages of development, members grow closer and bond with each other. Others will not and choose to leave. As your coven goes from being new and somewhat disorganized, perhaps with a "we'll make it up as we go along" attitude, to a more structured environment, as it must to survive, people do leave. The demands of their other life sometimes take people away. Being an active participant can demand too much time or commitment. If yours is a teaching coven, your members might begin the year as promising, challenging students, but drop out after a few months because they think that being a witch is too hard, or that the class commitment was too much. Cybercovens are easier to leave than physical covens; it just takes a click of the mouse.

Over the three years that I was a member of ShadowMoon, we had more than a hundred members, but only half of them actually stayed through their year and a day of training; fewer than that became participants in the core coven. Some left before Dedication, others before Initiation; some left in anger, others because their other life intervened; still others lost interest in the process. Wicca is not a religion for everyone, and many who are attracted to it leave once they realize the commitment required.

ShadowMoon had a particularly interesting problem with continuity. Our High Priestess spent most of a year being seriously ill and unable to be online for weeks at a time. Ritual healing helped; but then something else would happen. Luckily, during our first year her absences were not too bad and she was able to continue our lessons more or less on time. In the second year, however, it was extremely difficult. We created a High Court to provide structure during her increasingly frequent and lengthy absences, but the court was unable to carry the coven forward. We were paralyzed with indecision. But as we watched members leave as a result of our inaction, we learned from our mistakes and pulled the center of the circle in tighter to compensate. Our third year had the lowest membership, but was the most cohesive and creative.

Remember that, although you may join others on the path of Paganism, those other people chart their own course. Sometimes that means they decide to practice solitary or join another group. One of the most generous and loving acts a coven can perform is to say good-bye to a member who is leaving honorably. In a coven, losing a participant is truly like losing a family member, and other members experience anger and grief, as with a death. Depending on the reason for the member's departure, you can hold a ritual of farewell, or perhaps to just allow the coven to meet to say good-bye.

My formal training comes from a hierarchical tradition that uses three degrees to indicate knowledge and training on the pagan path. Although my High Priestess and I had discussed the inevitability of my leading my own coven after I received my second degree, I felt that my training was not up to the task. One night, I dreamed of my spirit guide, Grandmother Jaguar. She came to me in a midnight forest and simply said: "The name of the coven is JaguarMoon." The next day, I called my High Priestess and told her that I was now ready to take my third degree and hive off a daughter coven at the end of the year. I am not one to ignore the Goddess when She speaks!

The old laws, that is, Gardnerian and those traditions based on his teachings, say that "upon the instant" a coven member indicates a desire to hive, he or she is no longer a member of the mother coven. The reason for this law is a good one, in my experience.

Practically speaking, my High Priestess and I were aware of my plans for more than six months before we informed the rest of the coven. I was not the first witch in the coven to receive her third degree, but I was the first to hive. During my last months with the coven, particularly when no one else knew of my decision, I felt divided, stretched thin, and serving too many others. On one hand, I was a vitally important member of ShadowMoon, responsible for maintaining a smoothly running coven and engaged in discussions of coven policies. On the other, I felt pregnant with a new life, one that would demand much of my attention and that needed care and planning to be successful. It was a difficult time for me.

It was worse after the news was shared with the rest of the coven, because it was still many months before I would actually leave. In the meantime, we all began to go through a drawn-out grieving process—although, to paraphrase Monty Python, "I was not dead yet!" Several members felt I was abandoning them, others that I was skipping over them in terms of recognition, akin to being promoted over their heads. Turmoil resulted as issues were recognized and addresses, although not always solved.

Yet the birth of a daughter coven, much like the creation of the beehive (from which the term "hiving" is taken), is a vital to a coven's growth and continuity. I witnessed a proud moment in my original coven's growth when it became large enough and had trained members enough that a new coven was created. I felt called to celebrate the birth, despite the sharp pain of loss as the daughter or son leaves the coven. I chose to feel the joy at the accomplishments and a job well done. We marked the occasion with a special coven blessing, a ritual that empowers the new High Priestess or Priest to lead well and wisely, and offers a sense of closure for the community they leave behind.

A healthy, living coven or tradition needs a combination of continuity and change. Ideally, these elements exist in balance or in a flowing cycle, since continuity without change stagnates and change without continuity is chaos.

/part_three/

Practice

8: Preparing for Cyber Magick

Ritual preparation involves the physical creation and cleansing of sacred space, and mental groundwork and exercise. This chapter continues to explore how a cybercoven does magick online.

MENTAL PREPARATION

Ritual preparation involves training the most important tools of the Craft: the mind and the will. Whether comprised of newcomers to Paganism or long-time practicing witches, the group needs to learn to work magick together and direct the energies raised toward a specific purpose. The togetherness can be difficult to master. Creating togetherness generally comes through doing group exercises, physical contact, and sharing personal knowledge. These three things exercise the mind so that it can easily focus on a single concept, linking coven members to one another in an intimate, trusting bond to form the group mind.

Physical coven members can do things together outside the coven or get to know one another at scheduled meetings. They may find concerts to attend together, or plan picnics in the local park. They may make Saturday nights mandatory meeting times and then use that time to discuss readings, do group meditations, or even just breathe to raise energy.

Obviously, this type of physical sharing is not possible in a cybercoven. Instead, cybercovens create a group identity in three ways. We:

▸ Share the stories of our daily lives;

▸ Meditate;

▸ Visualize.

These practices are not so different from those of a physical coven; our contact, however, is made through ritual and daily communication on a list. Sharing stories, meditation, and visualization are so important to the healthy functioning of a cybercoven that the group should never stop these practices. Otherwise, life in a coven can become uncomfortable and awkward. Also, the sense of identification that flows from these disciplines is a crucial component to ritual work; it helps a diverse group focus on a single objective.

/Meditation: Stilling the Mind:

Witches meditate because meditation gets the mind into shape for ritual and magick. Most people are not aware of the amount of noise in their minds. This noise acts as static, interrupting the energy being raised during ritual. Meditation reduces the amount of static in the mind, producing an inner sense of quiet. This quiet state of mind signals the universe that a person is in balance; his or her inner and outer realities are acting in harmony. Meditation helps a person to transcend the noise of the chaotic mind, the stress of the mundane world, and get to the centered stillness of sacred space.

In a physical coven, meditation is taught in the earliest stages of participation. Guided meditations are generally performed at each meeting and the newcomer quickly sees their importance. Members are led through these meditations, taught how to meditate without guidance, and usually given a series of meditative exercises to perform daily on their own. As the newcomer masters the difficulty of sitting still, being aware but not thinking, focusing on either nothing or a single image, the difference in them becomes apparent to other coven members.

The methods used in a cybercoven are similar to those in a physical coven, except that self-reporting is the only way to gauge a member's ability. If JohnWinterBear says he is meditating daily, I have to take him at his word. He and I know that it is his loss if he does not. The discipline of daily meditation is an exercise in personal responsibility.

Cyber groups can use guided meditations in their rituals or during lessons, but visualization is a technique that is only useful for advanced members with several years of training. New members often find it difficult to achieve a trance state, follow directions, and yet not think during the process. The trance state is lighter online because more of the rational brain is engaged in reading the directions. Nevertheless, online meditations can be read and performed by the coven, and constant practice increases members' ability. As always, just try it, practice it, and enjoy the benefits of successful online meditations.

Many books deal with the subject of meditation in greater detail and background. Some basic rules for meditating include:

1. *Meditate daily.* Meditating daily opens your mind to silence. It gets you used to not thinking, to merely being aware, to being rather than becoming or going. I recommend starting with five minutes daily, then increasing your time gradually until you are able to meditate for a half hour at a sitting.

2. *Meditate at the same time each day.* By meditating at the same time, when you first awaken for example, you create a habit, increasing the likelihood that you will continue the practice daily.

3. *Meditate at the same place every day.* Create a sacred space, a corner of your home where no one else goes. Make this a place that becomes so imbued with the Divine that you feel calm when you enter. This space may be your altar or your workroom, or simply a pillow in a corner with a candle and an incense burner nearby.

4. *Track your meditation experiences.* It is heartening to see your progress when you meditate. Tracking also gets you into the habit of recording the intuitive impressions that you get as you advance in the Craft.

5. *Choose a form of meditation that works best for you.* There are many ways to meditate. If you are new to meditation, try several methods, each for a month or more, before deciding if it feels right. Even if you have been meditating for a long time, it can be fun to try a different method. At this time, there are 5.52 billion ways to meditate—one for each person on Earth.

/Meditation Methods:

There are four basic methods of meditation. Choose the one that works for you.

/FIRST METHOD:/

▸ Sit still and empty your mind, letting intrusion come and go, paying no attention.

▸ Place your hands on your knees. Tilt your head slightly upward. Place your tongue on your upper palate and breathe through your nose.

▸ Breathe from the diaphragm, which is the area just under your rib cage. (If you are new to this form of breathing, try placing a hand over that area so you can feel it expand on the intake and contract with the exhalation of each breath.)

▸ Follow your breath into your body, and then out again. Think of nothing else. Just follow the breath.

▸ If a thought intrudes, simply let it go. Do not focus on it; just turn your attention back to your breath.

/SECOND METHOD:/

▸ Sit still and focus on a single image or word, letting any other intrusion go, paying no attention to it.

▸ Sit in a chair or on a pillow, with your hands resting on your knees. Tilt your head slightly upward. Place your tongue on your upper palate and breathe through your nose.

▸ Breathe from the diaphragm, which is the area just under your rib cage. (If you are new to this form of breathing, try placing a hand over that area so you can feel it expand on the intake and contract with the exhalation of each breath.)

▸ Breathe deeply and follow your breath into your body and then out again, until you are calm and relaxed. Focus your attention on a mental image: a mandala (or pattern), a place, or a word. Alternatively, you could chant a syllable or phrase. Vowel sounds work well, as do the names of deities or the satisfying OM.

▸ Whether your focus is on a visual image or a vocal one, focus your attention completely. If you are distracted, simply let the distraction go and turn your attention back to your image or vocalization.

/THIRD METHOD:/

This form of meditation is based (very loosely) on the work of an amazing woman, Gabrielle Roth. If this meditation appeals to you, I highly recommend her book *Sweat Your Prayers* and her series of tapes and CDs for further development. In this practice, a piece of music is used that is (mostly) instrumental and includes a variety of rhythms.

▸ Move your body to music, paying attention only to your movements as you are performing them. Empty your mind.

▸ Choose music to get your blood flowing and your body warmed up, and then move: roll your spine, stretch your hands, arch your arms. Move for at least ten minutes until you are completely relaxed and warm; you should be sweating slightly.

▸ Spend five minutes moving to music that flows like a pebble dropped into water; imagine circles within circles.

▸ Spend five minutes moving to music that is staccato like fire, bright bursts of energy, sharp angles, and swift decisive movements.

▸ Spend five minutes moving to music that is chaotic: a mix of flowing and staccato.

▸ Spend five minutes moving to music that is lyrical; a pattern emerges from the chaos and you are lifted up and away.

▸ Spend five minutes moving to music that is still... quiet and slowing... movement that exists, but is not large.

While you are moving, your mind is focused only on your current actions, not what you will do next, or how you look, or anything else. Focus only upon what you are doing in the here and now. As you become accustomed to this style of meditation, you will be able to spend longer amounts of time in the rhythms. Vary the time that you spend in each tempo if you feel the need.

/Fourth Method:/

This meditation, sometimes called the "walking or running meditation," comes from the Vietnamese Buddhist monk, Thich Nhat Hanh.

▸ As you walk, breathe from your diaphragm, easily and without force;

▸ As you breathe in, say to yourself "Breathing in, I calm my body";

▸ As you exhale, say to yourself "Breathing out, I smile" (the form is important to the rhythm);

▸ Continue.

Meditation is a valuable tool for coping with stress and its symptoms. By taking the time to be still, you will begin to find inner peace and tranquility. That peace comes with you when you return to reality mode. It can also help you when you are feeling overwhelmed or confused. By meditating regularly, you will find clarity in the silence.

/Visualization and Creating Focus:

Yet another way to unify the group mind is to practice visualization, a skill so important that the basic techniques are taught at the earliest stages of almost every pagan tradition. In particular, Wicca uses these practices to teach the practitioner how to achieve higher levels of consciousness, the awareness to see other levels of reality, and to function on other planes of existence.

Heightened intuition, trance work, and increased psychic abilities are all indications of your progress in the Craft. When you advance enough to access other planes of reality, you can focus and direct energy; in other words, you can perform magick. The guiding principle of Wicca is to develop the spiritual and psychic abilities of each individual witch.

Visualization is the process of seeing mental images, much like daydreaming, except that you direct your mental focus toward particular images or patterns. Pagans must maintain proper focus on an idea to perform successful magick. Visualization builds on the discipline of meditation to achieve that focus. We visualize an apple in our hands; we feel its weight, see its varying colors, and smell its tart freshness. We bite that apple and taste the robust flavors bursting on our tongues. We chew. We swallow. We continue biting and chewing until the apple is gone. At first, the apple is merely an image: a red circle in the area of our palm. But as we continue to practice, to build the strength of our mind's eye, the circle becomes a red apple. Gradually, the apple takes on weight and hue, acquires scent, and so on, until we eat the apple and it fills our stomachs.

Maintaining this level of detail while visualizing is vital to the coven's energy, as well as to the development of a witch's ability. How can you transform reality if you cannot envision the end result as a reality? If I wish to find a new job, I begin the process by seeing myself walking into a new office, meeting new people, and hearing them say "Good morning, Lisa, welcome to your new job." Only then do I light a candle and ask the universe to help me out while I scan the newspaper for places to send my resume. In other words, if I have a specific goal I want to achieve, I begin with the end, as if it is already accomplished. Another example is this book you are reading. The dreaming and planning came a long time before the writing, but the writing was accomplished by a clear vision of the outcome. This book is a magickal document in every sense of the word.

In a cybercoven, visualization is one of the most important skills we must master in order to succeed. Because of the lack of physical cues, we must engage the mind and translate the text information into three-dimensional, life-like, sensually complete data. Without being able to "see" the ritual being performed, our experience of the event will be lacking the richness and depth of true performance.

In visualizations, we focus on symbols. Symbols surround us in our lives: red signs warn us to stop, crosses denote a wearer's religion, and shapes follow in a sequence that forms the sentence you are reading. Witches use symbols and the visualization of those symbols to transform reality through ritual. In ritual, every movement and word relates to the purpose of the ritual itself. The witch manipulates these symbols to engage the subconscious and draw it into harmony with the conscious mind, so that all aspects of the person are focused upon the goal of the ritual: producing the desired result.

When JaguarMoon is in circle, and I cry: "Mighty Mother, strike this blade with light!" we are all standing together, arms raised, watching as a beam of light comes down and reflects off the blade. The beam is real because we in the coven are visualizing the scene, and our focus merges into the same reality. In some of our rituals, we go through a greeting process in which we link to one another. The High Priestess turns to the coven member on her left and says, "I greet thee with a kiss and hold your hand in mine." We all watch as she kisses and then clasps the hand. The coven member then turns to the person on his or her left and repeats the process, and so on through the circle, until we are all linked. It does not matter that our watching is entirely mental, and that we know the action is happening because we are reading a screen. Our visualization is so concentrated that we see ourselves standing between the worlds, next to one another, regardless of physical distance.

Volumes have been written on the subject of visualization. Some visualization exercises follow that will get you started. Also, do research and experiment with creating your own visualization techniques.

/EXERCISE #1: WHERE ARE YOU?/

Get comfortable and relaxed. Open your eyes and look at the room in front of you. Look for about a minute and then close your eyes. Visually recreate the room on your eyelids. Try to get as detailed as possible. When you think you have it perfectly duplicated, open your eyes and check your internal image against the external.

Go to a different part of the house (not too far away, though) and look at a room. Pay attention to all the details. Return to your meditation area and get comfortable and relaxed. Mentally recreate the other room.

Do the previous exercises, adding other sensual descriptions of the rooms: smell, taste associations, sounds, and tactile memories.

/EXERCISE #2: EATING WELL/

Visualize a piece of fruit. Look at it from all angles. Feel its weight in your hand(s) and the texture of its surface. Smell it. Take a bite from it and taste the fruit's essence. Now what does it smell like? Chew the piece and feel its texture on your tongue. Swallow it. Repeat until you finish eating.

/EXERCISE #3: GOING WITHIN/

This visualization takes you inside yourself. Do it in steps at first; you will progress as you practice. Always take these exercises in small increments, rather than moving too quickly and potentially injuring yourself. If possible, have a partner read each step to you, pausing to allow you to concentrate and absorb each step. When you feel that you have reached the place fully, give a slight nod, or lift a finger to let your partner know that he or she can continue.

1. Find a quiet place, assume your meditation posture, breathe, and close your eyes.

2. Extend your senses and become aware of everything in a six-foot sphere around you. Become aware of the way the air feels, the sounds, the smells. This sphere is your whole world.

3. Begin to bring the sphere in closer, until it is just encompassing your self. Feel how your body feels—the temperature of the air on your skin, the texture of the clothes against your skin, the smell of your body or any scent you are wearing—notice even the tiniest details. You may begin to feel as if there are insects crawling all over you, and you need to itch worse than at any other time in you life. Ignore it. You mind does not like to concentrate that hard, so it begins to send false sensory signals to you in an attempt to distract your concentration. After a while, the itching sensation will eventually go away.

4. Focus your attention on a one-inch square of skin. Feel everything about it—the way the hair follicle feels as air passes over it, how warm or cold it is, the way any fabric on it feels. This inch of skin now is your whole world and nothing else exists. Get to know this feeling, until it seems that it has always been that acute and detailed.

5. Picture yourself standing on this patch of skin. You are so small that the hair seems to be like a tree trunk and the natural wrinkles in the skin are like big fissures in the ground.

6. Focus your attention even more tightly now and become smaller. See the skin cells packed tightly together. You are a giant, standing on this new world of cells. Concentrate on this feeling, remembering it and locking it into your mind.

7. Make yourself smaller. You are now standing on the surface of a single cell. You are able to see all of the parts of it, as if looking down into a

pool of still water. The other cells are simply a short jump across a narrow gap. Concentrate on this feeling; lock it into you mind.

8.   Now make yourself even smaller. The surface of this single cell is the size of the world to you. Other cells seem like the stars, they are so large and distant. Concentrate on this sensation, remembering it and locking it into your mind.

9.   Now make yourself so small that you fall into the cell, landing on a world of individual atoms. You are a giant, standing on miles and miles of solid atoms. You cannot see the whole of the cell anymore, only its millions and millions of atoms. Concentrate on this feeling; lock it into your mind.

10. Smaller still. You are now standing on hundreds of atoms. Concentrate on this sensation; lock it into your mind.

11. Shrinking again, you are now standing upon the shell of a single atom. Looking around, you can see other atoms not very far away. Concentrate on this feeling; lock it into your mind.

12. Now you are riding a single electron as it spins. You ride it like an adult on a child's bike. The nucleus of the atom seems as far away as the Moon now. Concentrate on this feeling; lock it into your mind.

13. You can now stand on the electron as it spins around a more distant nucleus. The nucleus seems to be as far away as distant mountains now. You can faintly make out the hills and valleys that are the positrons and neutrons in the nucleus. Concentrate on this feeling; lock it into your mind.

14. Now you have left the electron and are on the surface of the atom's nucleus. The positrons and neutrons are like sand, you are so much larger than they. Concentrate on this feeling, remembering it and locking it into your mind.

15. You have shrunk yet again. The surface of the nucleus is like hilly terrain. You walk on and through the hills and gullies of positrons and neutrons. Concentrate on this feeling; lock it into your mind.

16. The positrons and neutrons are mountains and the spaces between are

deep valleys. Look about, see and feel the surface of this place. Concentrate on this feeling; lock it into your mind.

17. You now shrink yourself once more and you are falling, floating in the space between the enormous bulks of the positrons and neutrons. It feels as if you are in a shower of meteors in space. Concentrate on this feeling; lock it into your mind.

18. You shrink yet again. The positrons and neutrons are now distant stars as you float in space. You feel warm and safe in this place. In this place, there is nothing but you and you are everything in this place. There is no outside world. There is no stress. There is no tension. There is only peace and solace.

19. Curl the tips of your ring fingers to touch the tips of your thumbs and touch the tip of your tongue to the back of your top front teeth. The touching of the fingertips and of the tongue to the teeth is a lock. In the future, you will only need to touch your fingertips together and touch your tongue to your teeth to go straight to this deep level of awareness.

20. Concentrate on these sensations. Make them part of one another; lock these sensations into your mind together.

21. Now imagine a light in the distance, brighter than what you left behind. You will begin to get smaller and move toward this light. It grows and grows, until you stand on it. Concentrate on this feeling; lock it into your mind.

22. You are now shrinking again and falling into the light. This light fills your being with wholeness. This wholeness connects you with every-thing in the universe and gives you a feeling of completeness that cheers and lifts your spirit to heights you have never felt before. Your whole being is filled with health and energy. Curl the tips of your ring fingers in to touch the tips of your thumbs, touching the tip of your tongue to the roof of your mouth just behind the teeth. Concentrate on this sensation; lock it into your mind.

23. As you close your meditation at the end of the various steps, tell your-self to count to three and come back to the world in a natural manner,

at an easy pace that makes you comfortable. You should feel more relaxed and at peace than when you began. Sudden, abrupt returns can cause physical stress and tension.[1]

/EXERCISE #4: GOLDEN LIGHT/

This exercise is good for sunny days. My cats always come and meditate with me when I do this one. And a cat's purr is soothing and lulling.

Sit in a relaxed position, back straight and chin lowered slightly, in natural sunlight. Offer no resistance to the light, but allow it to fill you up—entering at your lowered eyelids, flowing through your head, down your neck, over your shoulders... filling up your torso with warm light and clear energy.

Encourage this light to flow into you with each breath, bringing peace to the heart and a clear mind. Offer no resistance. Let the light circulate and flow through you.

When you feel it is time to end, gently let the light flow back out of you. Let it return to its source, leaving you feeling clean and refreshed. Try this exercise in the moonlight as well.

/EXERCISE #5: EMPTY BOWL/

Visualize your body as an empty bowl. Feel your inner space. Let your breath flow into the bowl, gently cleansing it. This bowl inside your body is infinite in size. Know that it can contain all the stars in the universe; in fact, it can contain anything that you wish.

Now, feel the Divine Presence that gave birth to the bowl. You are the space in the bowl, no different from the space without. Stay in this place as long as you want. When you are through, release the bowl back into space, and return to your body.

PHYSICAL PREPARATIONS

The physical aspects of ritual preparation are simple, compared with the mental preparation. Physical preparations vary from ritual to ritual, and coven to coven. In some covens, only certain members are allowed to prepare the ritual space; in others covens, the duties rotate through the membership. Some rituals require elaborate preparation with individual masks, costumes sewn, and makeup applied to create the dramatic effect of a sacred play. Creating sacred space is a vital component of a successful ritual. Proper preparation sets the stage and persuades the mind to let go of its rational, logical thoughts. Allowing the power of the creative mind to emerge is the first step in doing ritual and creating magic.

However, in a cybercoven, physical preparations are made individually; each member prepares his or her personal sacred space. First, most people physically cleanse their body. For those of you with bathtubs, bathing by candlelight is wonderful, but a ritual shower is just as good.

/Cleaning the Sacred Space:

We also clean our soon-to-be sacred space as well. Vacuum the rug or sweep the floor, visualizing the "dust" of negative or flat energies being sucked out of the ritual space into the vacuum cleaner, or pushed into the dust pan and discarded elsewhere. When you are finished, the room should have an energized and fresh feeling. Return the space to a neutral energy setting. Just prior to creating the circle, you can use a turkey feather to clean the space, or smudge the area with sage. Anything more is up to you. If you treat the cyber ritual as anything less than a sacred act, you will not be satisfied with the result; so take the time to prepare carefully.

/Setting Up the Cyber Altar:

In my class and coven, when preparing to do ritual, we transform our desk or other computer location into our altars. I literally disassemble my main altar and set it up again on my desk and then repeat the process in reverse, clearing the area until the next ritual. Others in the group have two altars: a main altar and second one for cyber rituals. Still others have an altar next to their desk; when they cast their personal circles, they include both.

How you set up your cyber altar depends on many factors: your desk size, what you feel is necessary for ritual, and how your desk is laid out. I can only make a few recommendations:

▸ Use nothing hot that will damage you or your electronic devices;

▸ Do not let smoke blow into the CPU;

▸ Ensure that you can easily type and move objects on your altar during the ritual.

If you burn a lot of candles, please remember that smoke can damage the performance of your CPU. You may burn candles somewhere other than your computer area.

My altar is shaped like an equal-armed L and fits perfectly into a corner. This allows me to set up my altar in the North or the East. In rituals where having a South or West altar is more appropriate, I construct a temporary altar in the corresponding area of the room. Because of its shape, I only place an altar cloth over one side, and just clear the other section of work to do or bills so that nothing will distract me. My cats like to lie there during my rituals, occasionally moving to my lap.

I have a friend who has constructed his altar next to his computer desk, and simply calls a circle large enough to encompass both. A member of JaguarMoon has her permanent altar in the adjoining room, against the same wall as her computer desk. She casts a circle through the wall. In the end, you decide what feels right.

With the construction of an altar and the appropriate mental preparation, you are now ready to participate in an online ritual and perform cyber magick.

HOW TO HAVE A GREAT CYBER RITUAL

As in physical covens, the ability to alter one's consciousness is a learned skill and individual experiences can be vastly different within the same ritual. Some members feel strong currents of energy flow through them throughout the entire ritual; others feel waves and surges of energy; still others feel nothing. This is one aspect in which the coven's training in visualization, focus, and concentration skills has the most impact.

Despite the arguments of people who do not believe in the efficacy of cyber rituals, my experiences show that rituals raise power, no matter where the coven is located. Although I sometimes feel less energy when in a cyber circle than I ever have in a physical ritual, it is nonetheless tangible. I have also found that, if I dismiss the ritual as not having raised energy, I experience problems with my physical and emotional balance for a time afterward. The energy raised in a cyber ritual seems to be subtler than that of a physical ritual, but is nonetheless present.

There are times when I am aware of sitting in a darkened room, staring at, essentially, a TV screen, waiting for something—anything!—to happen. Other times, I have felt energy flowing through me and my hair standing on end as the hands of the Lady touched my forehead in blessing. There has never been an occasion—physical or cyber—while interacting with the Lord or Lady when I was not acutely aware of Their presence within my circle.

The most important thing you can do to have a great cyber ritual is to take it seriously. Just because no one can see you does not mean you can spend your ritual time in front of your computer, eating Fritos, drinking a Coke and toggling between IRC and Solitaire!

Here are some specific steps you can take to have a great cyber ritual:

1. Take a ritual bath or shower, cleansing yourself of the day's residue. Say a blessing as the water flows over you. In a bath, put rosemary and salt into the water as it fills the tub. In the shower, a nice thing to do is to tie a bunch of fresh herbs to the showerhead. In any case, start your day over, fresh and clean.

2. Make sure your ritual time will be undisturbed. Just because you are physically "there" does not mean you are available to other members of your household. Be very clear and firm about this. I have a friend who has a "No fire, no blood, do not bother" rule for her household. When she is in ritual, unless something is on fire, or someone is bleeding, they do not disturb her. It seems to work very well.

3. Wear your ritual clothing. I actually have two sets of robes, and I choose between them according to the air temperature. Fuzzy cotton, a thin fleece for autumn, winter, and early spring, and then a lighter cotton robe for late spring and summer. I add a pair of socks in the winter because, unlike in a physical coven, it can get very cold sitting in one place.

4. Cast your personal circle around your altar and include your computer. See it as a blue or white sphere glowing on a map of the Earth.

5. During the ritual, visualize all actions as if they are including you. Watch the opening and closing of the circle, see the athame in the High Priestess' hand, linking together all of the members until the world is encircled in shining jewels of light. Sip from the chalice as it is passed to you, and so on.

6. If, during the ritual, you feel your attention wavering or your energy

flagging, take a moment to regroup. Human energy fluctuates. Pause and center yourself. Chant the words on the screen out loud. Meditate for a few moments on the purpose of the ritual. Ask the Lord or Lady for a "boost."

7. Remember to allow humor into the circle. The world works in mysterious ways; typos happen and some of them are truly hilarious.

8. Remember that, if you get "pinged out" or disconnected from IRC, remain calm. Log in again and re-enter the circle quietly. Cyber circles are extremely forgiving.

9: Ritual Examples, with Commentary

This chapter contains a variety of cyber rituals. The comments indicate the similarities and differences between physical and cyber rituals. I have transcribed an Esbat, a ritual including a guided meditation and a healing ritual.

If you are going to chant or work offline (like writing something) have everyone agree to a symbol (!, for example) that signifies when they have released the power, want to move on, or have fallen silent. This way, flow happens, and the leader knows where people are in terms of their energy.

NEW MOON RITUAL FOR PROSPERITY

Created by Ma'at for ShadowMoon Coven

Preparations: Charge a green or white candle by holding one end and stroking from the tip to the middle. Visualize prosperity and abundance flowing into the candle. Repeat with the other end of the candle. Assemble a piece of paper and with a writing implement, write a description of what poverty looks like to you. Is it a person? A smell? Be brief but clear about exactly what you think represents poverty.

Ritual:

High Priestess:

All Hail and Merry Meet JaguarMoon Coven!

We gather here in the dark of the Moon

to honor the Lord and Lady

and receive the bounty of Their gifts.

I conjure thee, O circle of art,

To hold us and protect us as we venture

between the worlds

Wherefore do I bless and consecrate thee...

So mote it be!

All reply: So mote it be!

High Priestess:

Black spirits and white,

Red spirits and gray,

Harken to the rune I say.

Four points of the Circle,

Weave the spell

East, South, West, and North, your tale tell.

East is red for the break of day,

South is white for the noontide hour,

In the West is twilight gray,

And North is black for the place of power.

Three times round the Circle's cast

Great Ones, Spirits from the past;

Witness and Guard it fast.

So mote it be!

All reply: So mote it be!

High Priestess:

I stand at the Eastern Gate.

Eastern Guide:

Spirits of the East, we do summon thee

tonight!

One of the nuisances of IRC is that if you do not type anything for a period of time (as little as a minute, as long as 15 minutes) you get "pinged" out, or disconnected. So we try to add "responses" throughout the body of the ritual to keep people connected.

Notice how the imagery used by the High Priestess in the Casting of the Circle is so explanatory. You know exactly where she is and what she is doing. As a result, the entire coven is able to easily visualize her actions, producing a group mind.

O Powers of Air, Givers of Wishes,
O Bringers of Dreams, Protectors of Magick,
Keep us safe within thy sight,
Protect this Circle with thy might!
So mote it be!

All reply: So mote it be!

High Priestess:
I stand at the Southern Gate.

Southern Guide:
Spirits of South, we call to thee!
Forces of Fire, Bringers of Change,
Sweet blood of life, Passion's wild range!
Our magick, keep pure, our flame, coax bright;
Guard us, our circle, on this winter's night!
So mote it be!

All reply: So mote it be!

High Priestess:
I stand at the Western Gate.

Western Guide:
Spirits of the West, we do summon thee
    tonight!
O Powers of Water, Givers of Love and
    Friendship,
O Keepers of Karma, Protectors of the Gates
    of Death,
Keep us safe within thy sight,
Protect this Circle with thy might!
So mote it be!

The theme of the Quarter Calls includes gifts, change, and protection—appropriate to the ritual.

All reply: So mote it be!

High Priestess:
I stand at the Northern Gate.

Northern Guide:
> Spirits of North, we call to thee!
> Forces of Earth, Essence of Stability,
> Keepers of Wisdom, Grace of Fertility...
> Our magick, keep wise, our truth, hold right;
> Guard us, our circle, on this winter's night!
> So mote it be!

All reply: So mote it be!

High Priestess:
> Heavenly Goddess!
> Your breath gives us precious life and creation.
> As we stand here before you, a strong growing
>   nation,
> Earthly Lady!
> Your body bears the fruit of your children.
> And to you, we return, the circle unburdened.
> Oceanic Mistress!
> Your waters run deep through our young,
>   seeking minds,
> We ask for your blessings, our magick's true
>   bind,
> Fiery Enchantress!
> Your blood pounds the fury of life through our
>   veins.
> We honor your aspects, your waxes and wanes.
> We seek you this night, in the darkest of
>   hours.
> Our coven, you grace, with your knowledge
>   and powers.
> Mother, yet Maiden, yet Crone do we see...
> And, as we call, so shall it be!

High Priest:
> Solar God!
> Your warmth brings forth young shoots from
>   the earth.

We honor your guidance, passion, and mirth.
Hunting Master!
Your beasts roam the wilds in the great circle
    game;
Accepting their sacrifice, we feast in your
    name.
Protective Consort!
The throb of desire, the lust in your loins;
You ensure our survival when your lover
    you join.
Loving Child!
From the pit of darkness, your wailing was
    hailed!
As the bringer of light, cold winter unveiled.
We call you this night, as Guard, Lover and
    Son,
To protect this, our coven, and magicks
    we've done.
Ensure our endurance for the circle you see,
And as we call, so shall it be!

High Priestess:
    The circle is complete;
    The circle is sealed.

The four Guides:
    N: What is this night?
    S: It is the night of the dark of the Moon.
    E: What is the meaning of this night?
    W: It is a peak of power.
    N: What is the element that rules this night?
    S: Darkness rules this night.
    E: After tonight, what power will wane?
    W: Darkness will wane and light grow.
    N: How do we recognize ourselves tonight?
    S: We greet the Moon and glory in the darkness.
    E: Who helps us?
    W: The Lord and Lady help us.

This section should be spoken (that is, typed) by each of the four Guides in turn, creating a kind of antiphonal chant.

N: What is our God?

S: He is the wild places, and the dark alleys, the Hunter, the Lover, the Sage.

E: What is our Goddess?

W: She is the cold of space, the deepest abyss, the Crone, the Virgin, the Mother.

N: Where are the God and Goddess?

S: They are in our hearts.

E: Who are they?

W: Behold, they are ourselves.

High Priestess:

Participants take out their poverty paper.

There is a place inside all of us where poverty lives.

Take a deep breath

in....

out...

in....

out....

in again....

out....

Go to that place and look at Poverty

Look at your list...

What does being poor/Poverty

Look like to you?

Smell like?

Taste like?

Sound like?

Feel like?

Focus on the essence of being poor...

What is it?

It is merely a state of being that can be changed.

You need not suffer poverty...

How does it make you feel?

Think about:

your anger

This begins the "interactive" portion of the ritual. Notice how this reads like a physical ritual. As in guided meditations, there are pauses between each phrase, allowing the participants to focus on the concepts they hold of poverty.

your hopelessness
your fear
shame
sadness
Now...
Destroy your paper by wadding it into a ball
and place it into your cauldron.
Light a fire of release!
As the paper burns, see the poverty flow from
you.
Burn the paper completely to ashes.
Repeat this mantra several times as your
poverty goes up in smoke:
Poverty flow from me
As I will, so mote it be!

All reply: So mote it be!

High Priestess:

  Now light the green candle.
As you stare into its flame, visualize money
flowing into your life.
See yourself, handling the denominations.
Pay your bills; when you are done, there is a
positive balance in your checkbook.
See the balance growing larger.
Spend the money on things you want and
need.
We give ourselves prosperity as we raise the
energies
Within this magick circle.
Send that energy out into the universe to do
your bidding.
And, once it leaves, never speak of this ritual
or you risk fettering the energy.
Breathe easily now... the negative is gone.
The positive remains.
We are wealthy.

This part of the ritual "feels" like a physical ritual; in fact, it essentially is the same.

High Priest:

> Thank you Lord and Lady for Your bounty and
> > Your grace.
> The abundance You weave for us is accepted
> > gratefully into our lives
> To nourish and sustain us as we work Your
> > will in the world.

High Priestess:

> I stand at the Northern Gate.

The theme of the
Quarter Calls includes
gifts, change, and
protection—appropriate
to the ritual.

Northern Guide:

> Powers of North, of Wisdom,
> Thank you for your protection and blessings
> > this night....
> Hail and farewell and thank you...
> Blessed be!

All reply: Blessed be!

High Priestess:

> I stand at the Western Gate.

Western Guide:

> We thank thee, Spirits of the West,
> For guarding us while on this quest.
> We send thee now unto thy rest!
> So mote it be!

All reply: So mote it be!

High Priestess:

> I stand at the Southern Gate

Southern Guide:

> Powers of South, of Life's Flame,
> Thank you for your protection and blessings
> > this night...

Hail and farewell and thank you...
Blessed be!

All reply: Blessed be!

High Priestess:
I stand at the Eastern Gate.

Eastern Guide:
We thank thee, Spirits of the East,
For safety during our spiritual feast.
We send thee now to go in peace!
So mote it be!

All reply: So mote it be!

High Priestess:
Please ground your excess energies in Mother
Earth;
Feel them flowing into Her as you place your
palms on the ground.
The Rite is complete.
May the Circle be opened, but never broken.
May the Peace of the God and Goddess be ever
in your hearts.
Merry meet, merry part, and merry meet
again!
So mote it be!

This Prosperity ritual was somewhat unusual, in that
prosperity rituals are generally performed at the waxing Moon and are symbolic of wealth flowing into
your life. I created this ritual after working with a
wonderful book by Timothy Roderick called *Dark
Moon Mysteries*, which includes a prosperity ritual.
His rite was the genesis of mine, as I was struck by the
idea of casting poverty out of my life. Before we can
draw new things into our lives, we must clean out the

old and unwanted—like cleaning out your closet to make room for new clothes.

A HEALING RITUAL

Created by Aurian for ShadowMoon Coven

Preparation: Light a white candle just prior to ritual. Burn sage or sandalwood incense.

Ritual:

High Priest:
Be it known that our Circle is about to be cast!
Let those who desire attendance at this healing
Gather in the East and await the summons.
Let none be here other then by their own free
    will.

The High Priestess casts the Circle.

East Caller:
All Hail the Guardians of the East,
Spirits of Air who come on
Wings of Healing, the wind of understanding,
Carry us aloft into the clear blue sky,
Inquisitive and curious, awakening our
    intuition,
Wise and knowing, that I might know why.
Welcome and blessed be!

These Calls invoke the powerful and healing aspects of each element.

All reply: Welcome and blessed be!

South Caller:
All Hail the Guardians of the South,
Spirits of Fire who burn in the

Flames of healing, sweet determination
Powering our belonging, making us strong,
Wild and hungry, joyful and insistent.
Home to desire, for the fire I long.
Welcome and blessed be!

All reply: Welcome and blessed be!

West Caller:
All Hail the Guardians of the West,
Spirits of Water, who swim in the
Waves of Healing, serpents of Mystery,
Swallow us below to the tender deep,
Strange and dreaming, wakening our wonder.
Home to the truth, that my heart not sleep.
Welcome and Blessed be!

All reply: Welcome and Blessed be!

North Caller:
All Hail the Guardians of the North,
Spirits of Earth, who live within the
Cave of healing, rhythm of our heartbeat,
Cradle us to silence, still the pain,
Cool and patient, welcoming and wordless,
Home in our bones, make us whole again.
Welcome and blessed be!

All reply: Welcome and blessed be!

High Priestess:
Earth, Fire, Water, and Air,
Bring your healing energies this night and
    grant us your blessings.
The Circle is cast;
We are between the worlds,
Beyond the bounds of time,
Where night and day, birth and death, joy and
    sorrow

Meet as one.
The Ritual may begin!

Great Mother of healing, be with us in this
    ritual of healing for [insert name].
Guide us to know what to do in helping [him
    or her] heal;
Help [him or her] to recognize and accept what
    [he or she] needs for healing.
We ask this in love.
So mote it be!

Coven Healer (or High Priestess):
    [insert name], please repeat the following
        affirmation after me:
    I, [insert name] am now willing to release the
        need for pain in my life.
    I release it now, accepting and trusting in the
    Lord and Lady and the process of life to assist
        me
    To heal and to meet all my needs and desires
        in a healthy way.
    I am a whole and perfect [son or daughter] of
        the Lord and Lady.
    So mote it be!

In this ritual, only one of the three people being healed was present. In your ritual, if the person cannot be present, you might have them meditate upon this affirmation during the time the ritual is being performed.

To the coven:
Focus now on your representation of [insert
    name].
[Insert name] focuses on [himself or herself]
    and in perfect love
And perfect trust accepts the healing that is
    being sent to [him or her].
Build your mental picture of [insert name];
Reach out to [him or her] through time and
    space to connect with
[his or her] Higher Self.

Let me know when you are ready to continue.

When ready, all reply: Ready.

Coven Healer (or High Priestess):
  Now that you have connected with [his or her]
    Higher Self,
  Ask to see what areas of [his or her]
    physical/auric body need to be healed.
  You may see holes or muddy spots in the
    auras,
  Or see hot spots on the physical bodies,
  Or just know intuitively which parts need to
    be healed.
  Mentally scan [his or her] body from head
    to toe...
  Let me know when you are ready to continue.

The ability to visualize comes strongly into focus during this aspect of the ritual.

When ready, all reply: Ready.

Coven Healer (or High Priestess):
  Now, ask for a color to send for healing.
  Using that color, fill their auras with the color.
  If it helps, use your breath to gently blow the
    healing color toward
  their auras...
  Let me know when you are ready to continue.

When ready, all reply: Ready.

Coven Healer (or High Priestess):
  Some parts of their physical body will need to
    be healed;
  Focus on these areas now.
  You will know intuitively what color to send.
  If it helps, gently blow the color onto the
    physical areas that need healing...
  Let me know when you are ready to continue.

When ready, all reply: Ready.

Coven Healer (or High Priestess):
 As a group, we will now surround [insert
  name]'s aura with the color green.
 Visualize [insert name] in perfect health,
 See [him or her] better than ever.
 Then imagine [him or her] five years from
  now, happy and in perfect health.
 Surround them with green again...
 Let me know when you are ready to continue.

When ready, all reply: Ready.

Coven Healer (or High Priestess):
 Take your representation of [insert name] and
  stand it upright.
 This is a symbol of [insert name] no longer
  lying low, but in good health again.
 Now, surround yourselves with light.
 Do not dwell on [insert name] or [his or her]
  illness,
 Or what happened during the healing.
 The rite is finished.

The High Priest walks to the center of the circle.

High Priest:
 Mother,
 Maiden,
 Crone,
 She who brings healing, love, warmth, and
  well-being,
 We thank you for attending this healing.
 As this Circle is dissolved,
 We ask that you continue healing in each of
  our

Because of the advanced level of skill required for visualization in this ritual, I would not perform it too early in a cybergroup's formation.

Lives and inspire us with the fresh creativity
   that you bring to all life.

North Caller:
   All Hail the Guardians of the North,
   Spirits of the Earth,
   Keepers of our flesh and bone.
   Thank you for your attention to our plea.
   Go if you will, you are free to roam.
   Merry part and blessed be!

All reply: Blessed be!

West Caller:
   All Hail the Guardians of the West,
   Spirits of the Sea,
   Protectors of our emotions and desires.
   Thank you for your attention to our plea.
   Go if you will, if you require.
   Merry part and blessed be

All reply: Blessed be!

South Caller:
   All Hail the Guardians of the South,
   Spirits of Fire,
   Shield of our will.
   Thank you for your attention to our plea.
   Go if you must, if it is your will.
   Merry part and blessed be!

All reply: Blessed be!

East Caller:
   All Hail the Guardians of the East,
   Spirits of Air,
   Paladin of our intuition.
   Thank you for helping us with our plea.

Again, notice how the Calls and Dismissals are different in each ritual, reflecting both the author, and the ritual's changing focus.

Although no specific deities are called upon in this ritual, you might call Diana/Artemis, Leto, Isis, Zeus, or Hephaestus.

Go if you wish, if it is your intention.
Merry part and blessed be!

All reply: Blessed be!

High Priestess:
The Ritual of Healing is complete.
May the Circle be opened but never broken.
May the Peace and Healing of the Goddess be
    ever in our hearts,
And may we spread that peace and healing
    wherever we walk.
Merry meet, merry part, and merry meet
    again.
So mote it be!

RITE OF MIDSUMMER (LITHA)

(as an example of a Mystery ritual)

Preparation: Place baskets of flowers at each of the four Quarters. Place masks representing the Maiden, Mother, Crone, the Young King, and Declining King around your altar. Position an unlit candle near each one.

Cast of Characters: Narrator, Maiden, Mother, Crone, Young King, Declining King,

Ritual:

High Priest:
Be it known that our Circle is about to be cast!
Let those who desire attendance at this
    Midsummer Rite
Gather in the East and be greeted!

Let none be here other than by their own
    free will,
in Perfect Love and Perfect Trust.

High Priestess:
    Mighty Mother!
    Strike this blade with light
    That I may cast the Magick Circle!
    I conjure Ye, O Circle of Art, to be a Temple
        between the Worlds,
    Wherefore do I bless and consecrate thee.
    Black Spirits and White,
    Red Spirits and Grey,
    Come ye, come ye, come ye as may!
    Around and around, throughout and about
    The Good comes in. The Ill keep out!

This is the "traditional" circle casting used by ShadowMoon. Notice how rich it is in imagery, making it easy for participants to follow the High Priestess's actions.

    I stand at the Eastern Gate:
    Hail Winged Ones of the East,
    Bringers of dawn,
    Bring us thoughts that soar among the birds,
    Skies swept clean by the winds of insight.

    I stand at the Southern Gate:
    Hail Bright Ones of the South,
    Lords of Noonday,
    Dance the fires of your passions within us,
    Set us ablaze with the summer's heat!

    I stand at the Western Gate:
    Hail, Laughing Ones of the West,
    Sunset waters,
    Bring sweet delight to those who thirst,
    Cleanse our Spirits,
    Wash our cares away.

    I stand at the Northern Gate:
    Hail, ancient ones of the North,

Trees deep-rooted,
Bring forth beast and bird, forest and field,
Let the sweet Earth turn beneath us as we
    dance!

High Priestess:
A ring of pure and endless light
Bounds this meeting place of souls.
Our energy and love unite,
Will serve to make this Circle whole.
Our presence sanctifies this place:
Time is not the master here.
We invite joy and love to abide,
And forbid unwanted spirits to appear.
The Circle is cast...
We are between the worlds,
Beyond the bounds of time,
Where night and day, birth and death, joy
    and sorrow
Meet as one.
The rite may begin!

Narrator:
Litha, also called Midsummer or the Summer
Solstice, celebrates the abundance and beauty
of the Earth. This is the longest day of the
year. From this day on, the days will wane,
growing shorter and shorter until Yule. The
trees and fields are prosperous. The young
animals and birds frolic in the fields and trees.
This is a time of the Faery. It is believed that,
on twilight of this day, the portals between the
worlds open and faery folk pass into our
world. Welcome them on this day and they
may bless you with their wisdom and joy! This
is a time to look internally at the seeds you
have planted that should be in full bloom.

*The scene is set, and
participants are given
the context of the ritual.
The "seed" reference
comes from the Imbolc
ritual in which we
planted the seeds of our
desires for the coming
year.*

High Priest:

    Great One of Heaven, power of the Sun,

    We give honor to you,

    And call upon you in your Ancient Names:

    Michael, Balin, Phol, Hugh, Herne, Heimdul,

    Balder, Arthur, Perkunis.

    We call upon you,

    A friend of times far ancient!

    Blessed be the God!

High Priestess:

    Great Ones of the Stars, spinner of Fates,

    We give honor unto you

    And call upon you in your ancient names:

    Fortuna, Aphrodite, Huldana, Mari, Freya,

        Morrigan, Cerridwen, Hecate.

    We call upon you,

    Friend of Times far ancient!

    Blessed be the Goddess!

Narrator:

    We have invited the Lord and Lady to join us

        in our Rite of Midsummer

    And their presence is here! We look around

        the Magick Circle;

    Our High Priestess lights a candle and holds it

        aloft before a mask.

    It is the Mask of the Maiden, and She speaks!

The masks reflect an aspect of the Deity, and each imparts its lesson to the coven.

Maiden:

    Life springs eternal, with joy and with beauty.

    Now is the time of celebration

    Of the powers of life and of sensuousness.

    Of newness and of richness,

    For richness and plenty do bless all the lands.

    Now shall be crowned the shining, sacred

        King,

In the full knowledge that
All that is good and pleasing
Shall ever return, as glorious as before,
As certainly as the coming of the New Moon.

Narrator:

As our High Priestess lowers her candle, we
look about the Magick Circle again. Our High
Priest lights a candle and holds it aloft before
another mask. It is the Mask of the Young
King ... He speaks!

Young King:

There is a time for all things to grow,
A time to build, and a time to explore,
A time for building empires,
And a time for laughing at the storm,
A time for exultation for the glory of the
      world,
And joy in one's own strength.
This is a time for celebration
Of the virile and bold season.
Now, with gladness, shall be crowned
The beauteous Sacred Queen,
In the full knowledge that times of joy
Shall return as always, lusty as before,
As certainly as the Sun shall rise again.

You begin to see the flow of the ritual. The High Priestess paces around the temple, holding her candle before each mask.

Narrator:

As our High Priest lowers his Candle we peer
in the dark around the Magick Circle. Our
High Priestess lights a third candle and holds
it aloft before another mask. It is the Mask of
the Mother and She speaks!

Mother:

As surely as the Moon becomes full,

All must ripen at last,
Seek their destiny,
And decline.
Here it is that
The proud banners are planted.
The rich cities shine in their heights of
    splendor.

One at last sees and fully knows The World,
The Gods, and one's own soul.
Whatever thence happens, or however,
Be it of a vast nation, or within
The breast of one alone and solitary,
A point has been made
Before Eternity itself
And, so shall it remain forever.

Participants hold each mask, or look deeply into its eyes as they "listen" to the lesson given.

Narrator:

As the High Priestess lowers her Candle, our
High Priest directs us to a fourth mask, now
illuminated in the Candle he lights. We see the
Mask of the Declining King...and He speaks!

Declining King:

The Sun must set
And its light fade.
In the scheme of all things,
There is a place for decline,
A time for ending affairs,
For the closing of doors,
For the razing of cities,
And the breaking of monuments.
For how can there be new life
Unless death makes clear the way
And removes that which is old and outworn?
Know well that ending is as necessary
In all things as is beginning,

And that there must be death
That ye may in time become as Gods
And, in your Quest,
Gain life eternal!

Narrator:

Our High Priest lowers his Candle and we look
about the Magick Circle one last time. The
High Priestess lights the last candle and holds
it aloft before the fifth mask. It is the Mask of
the Crone...and She speaks!

Crone:

The Moon must decline, darken, and die.
Beyond life are the realms of Shadow:
The journeys unknown and the visions
    undreamed.
For the soul seeks its own level, rests,
And gains a time and a place to be reborn
    again.

Yet, here also, between the worlds,
Is wisdom deep and power great,
That those who are fitting
In mind and in spirit may learn.
Ye who are worthy, seek this chalice of
    intellect,
Of attainment, of Magick.
Use it for your good, and the good of others.
For so shall ye gain, in time,
The power and wisdom of the Gods.

Narrator:

As the final Candle is lowered, and the mask
fades into the dark, we contemplate the words
that have been spoken by these Oracles. And,
we listen to those who guide us.

Our High Priestess and High Priest move to
stand before the cauldron on the altar. As she
tosses a handful of incense on the charcoal
and the smoke begins to rise, our thoughts,
like the sweet wafting scent, soar to new
heights and we beseech the Lady to help us
understand the lessons taught this night.

High Priest:
We give thanks to the God for knowledge of
Times to build, and times to tear down,

Power of will and strength of soul
Turn life's great wheel
And make us ever better
From life, to life, to life
Blessed be!

High Priestess:
That which we have seen and heard, regard it
    well.
There are turning points in all things,
Not only of this world, but others,
In the affairs of humankind
And with the Gods themselves.
Thus it is, and shall be,
To the ends of the universe and beyond Time.
For this wisdom,
We thank the Goddess.
Blessed be!

High Priestess:
I stand at the Northern Gate:
Farewell, Powers of the North!
Go in Peace to those mountain realms
from whence Ye came...
Hail and farewell, and blessed be!

I stand at the Western Gate:
Farewell, Powers of the West!
Go in peace to those ocean realms
from whence Ye came...
Hail and farewell, and blessed be!

I stand at the Southern Gate:
Farewell, Powers of the South!
Go in peace to those fiery realms
from whence Ye came...
Hail and farewell, and blessed be!

I stand at the Eastern Gate:
Farewell, Powers of the East!
Go in peace to those high and airy realms
from whence Ye came...
Hail and farewell, and blessed be!

High Priestess:

All Spirits whom this Circle has drawn,
We thank you and ask that you now be gone.
The Witches Law has always been,
You must go out where you enter in.
Hail and farewell, and blessed be!

This Rite of Midsummer is now complete.
May the Circle be opened but never be broken.
May the peace of the Goddess be ever in your
    hearts.
Merry meet, merry part, merry meet again.
Blessed be!

Conclusion

Pagans have worshipped their deities since the beginning of time, long before there were names for the religion. As culture brought new ideas, techniques, and tools, we adapted them to our use and profited from the knowledge.

Temples, groves, or covens probably came about as an efficient method of transmitting the sacred knowledge of the gods and goddesses. Sacred groups are still an effective method of transmitting knowledge, and online communication has made it more so.

We rush forward into the future with breathtaking speed, barely pausing to eat, make love, or sleep; so to talk of deities in the age of technology may seem strange. But this is not the case. More and more people are searching for meaning beyond the material world. A recent article noted that the so-called Generation X (born between 1961 and 1981), which has been labeled as lacking an interest in the spiritual, has actually been turning more and more to alternative religions. The Baby Boomers (born between 1945 and 1961) have also begun to look for spiritual, instead of financial, fulfillment.

Paganism is an old religion, but it is an adaptable one, constantly being recreated and developed, in defiance of absolutes and strict laws. Today, our symbols are drawn from nature, and we send them out over the Internet, sharing our knowledge with one another in glorious abundance. We watch as the lines dividing us become lines connected, allowing those who never saw unity to realize our essential sameness, unblurred by color, sexuality, economics, or geography.

The Virtual Pagan has been a pleasure for me to create, and I offer it to you, my reader, in the knowledge that what I have done, you will build upon.

And so our knowledge grows, in love,

LISA MCSHERRY
Lady Ma'at, High Priestess
JaguarMoon Coven

Appendices

A: Organizing Your Book of Shadows

You may have gathered from reading this book that I am a techno-geek, always looking to get closer to the cutting edge of technology. While this is somewhat true, it is not the whole truth. I am always looking for a better way to be organized. This outlook encompasses every area of my life, including my Book of Shadows.

WHAT IS A BOOK OF SHADOWS?

A Book of Shadows (BoS) is a record of magickal activities kept in some form or another by most occult practitioners. Some covens also keep a coven BoS. If so, then it will usually be accessible to all members. Students may be expected to hand-copy the coven's BoS into his or her personal BoS. In the latter case, the coven information is only a small part of the entire text.

The largest part of your BoS is the journal section. After every meditation, ritual, casting of a spell, or any magickal act, write down your feelings, observations, successes, and failures. Over time, this journal will grow increasingly valuable, as it charts your progress in the Craft. When you write your own rituals, discover new recipes, and explore new magickal lore, record it in your BoS. As you log your experiences at rituals and other magickal workings that you've attended, you can note what you thought worked well and what didn't. These notes will become valuable as you start writing and leading your own rituals. It will house clippings and notes, photocopies and photographs, and other bits of information that you accumulate. Do keep it organized.

How?

In creating my BoS, I began by sorting the materials I had into categories and filing my information. Included were such items as rituals and spells, recipes, writings, invocations, laws, herbs, Sabbats, and other magickal subjects. Originally, I had intended to copy them all into one book but ran into two impracticalities: 1) It was a lot of material to write by hand; and 2) If I had something to add to a category, it would be out of sequence, once the original entry was written in the book.

Then I decided to buy a journal book for each category, but realized that it would soon grow unwieldy. (Although I know of at least one practitioner who has a three-ring binder for each subject, I cannot stand the idea of a wall of binders—that is way too much information!)

I put most of the shorter pieces onto my computer. What was left was placed in a plastic sleeve (to protect it) and filed in my binder. Obviously, I add to the three journals on a regular basis. I buy more blank books as I fill the old ones.

I have a Tarot reading journal to track the readings I have done and gauge their outcome. I usually note such things as type of spread, time of day, deck used, date, and any influencing circumstances. I copy down the spread as it is revealed and may include some notes about my interpretation. It also documents my advancing skills and clues for getting more accurate readings.

My dream journal is one I write in nearly every morning. For me, telling the story of my dream helps me to clarify my insights. If my dreams have repetitive themes or images, I note that down. I also note my reaction to the images, my responses, and whatever seems to be relevant.

The personal journal contains my daily activities: descriptions of weather, events, thoughts, and feelings. Most of my activities are influenced by my spirituality, so it is also a good record of my progress and changing attitudes.

/Organizing Material for Your Book of Shadows:

There are two ways to look at keeping a physical Book of Shadows. In one, you view the process of collecting data as a record of a journey, with any organization coming from the progress of your knowledge and understanding. This is a very traditional way to use one's Book of Shadows. The other way is to create an expandable system to catalog and store the information you acquire. In this way, you are creating more of a filing method than a travelogue, perhaps by using binders or a filing cabinet, to store your data. One way to organize such a system is listed below:

Animals
Astral Projection
Astrology
   Chinese
   Other Systems
Auras
Basic Information
   General Terms
   Names
Being an HPS
   Counseling
Candles
Celtic Info
Chakra
Correspondences
   Colors
   Elements
   Tree
Crowley
Crystals & Stones
Deity
   God
   Goddess
Divination
Dreaming
Egypt
Energy
Ethics, Rede, & Laws
   Ethics
   Laws
   Rede
Food
Greek
Healing

Chinese
Eastern
Herbs
History of Wicca
   Founders
   Lessons
Humor
   Inspiration
   Myths
   Paganism
   Poetry
   Practical
   Quotes
   Stories
Initiatory Path
   Descent Myths for Second
   First Degree Stuff
Invocations & Songs
   Charges
Karma
Meditation
   Meditations
Miscellaneous
   Angels
   Consciousness
   Faery
Moon Magick
Native American
Numerology
Online
Other Traditions
   Druid
   Strega
Qabala & Tantra

Rituals & Spells
  Bathing
  Calling Quarters
  Circle Casting
  Esbats
    Dark
    Full
    New
  Healing
  House
  Humorous
  Love
  Miscellaneous
  Prosperity
  Protection
  Rites of Passage
  Ritual Components
Runes
Spirit World
Tarot
Teaching
Things to Make
  Bathing
  Crafts
Formularies
  Incense
  Oils
Tools
  Altar
  Athame
  BoS and Other Journals
  Chalice
  Pentacle
  Wand
Visualization
Wheel of the Year
  Beltaine
  Imbolc
  Lammas
  Litha
  Mabon
  Ostara
  Samhain
  Yule
Writings
  Essays
  Ezines
Yoga

/Adding New Technologies—
Creating an Electronic Book of Shadows:

A friend of mine writes in his BoS, and then later transcribes it onto the computer. He says that now he has it all organized in a way that makes sense to him.

One issue that I found extremely frustrating about keeping a physical Book of Shadows was the maintenance of all the information I was collecting from the Internet. I struggled with it for about a year and then gave up and converted the bulk of my files into an electronic Book of Shadows.

The Hard Disk of Shadows is a great option for storing huge amounts of information. You can save a few trees, prevent writer's cramp, and print out copies of relevant information or ritual scripts quickly. Also, you can quickly copy and paste online text into a script and then print it out. It is especially convenient to be able to print copies of a ritual in a font large enough to read by candlelight. (This brings up another point—the calligraphic font. Yes, calligraphy looks wonderful and makes impressive documents; however, it is nearly impossible to read in dim light. Choose a font that is easy to read and generic. No point in keeping your BoS in a fancy copyrighted font when it will appear as gibberish on a computer screen.)

I have a computer with two hard drives. The main drive (c:/) is where I keep all of my programs and personal information. The second drive (d:/) is entirely reserved for my occult workings. My HDoS is there, along with all information related to my coven, and any graphics that I saved.

My HDoS is structured so that the Book of Shadows directory is only one away from the root directory (d:/). Opening the Book of Shadows produces a collection of folders (Animals, Astral Projection, Writings, and Yoga, among others). Some folders are subdivided (Miscellaneous, for example, has Angels, Consciousness, and Faery as subfolders); others stand alone. I used the "Organizing Material" outline above to create the file directory structure.

/An Outline of an Online Book of Shadows:

I belong to several e-mail lists whose participants share information daily, and I find that copying the data, pasting it into a blank document, and then saving that text as a single file is the best method for me to organize information. I do this daily, saving interesting or unique information into a single folder. I pay close attention to how I name the file, because the title must tell me what information is contained therein, but not duplicate an existing filename. (Although I do have four files named Litha. I will need to consolidate and rename them soon.) Monthly, or more frequently if time permits, I move those files into their appropriate folders. Another way to organize your electronic Book of Shadows is listed below:

Book Blessing
Table of Contents
Charges
    Charge of the Goddess—Doreen Valiente
    Charge of the Goddess—Unknown
    Charge of the Dark Goddess—Tiger Eye

Sabbats and Full Moons for the Current Year
    Lunar Correspondences
        Astral Correspondences
    Celtic Lunar Year
    Celtic Trees and Their Months
    Celtic Birds
Sabbats, Esbats, and Holidays
    Samhain
    Yule
    Imbolg
    Ostara
    Beltaine
    Midsummer
    Lugnasadh
    Mabon
Ceremonies
    Self-Blessing
    Bathing and Self-Blessing Ritual
    Cleansing Ritual
    Birthday Spell
    Naming
    Dedication Pledge
    Self-Dedication Ritual
    Initiation
    Handfasting
    Handparting
    Wiccanning
    Maiden—Mother
        Croning Ceremony
Gods, Goddesses, and Pantheons
    Gods and Goddesses by Ariadne
    Hellenic
    Celtic
    Norse
    Egyptian
    Sumerian
        Hindu
    Gods

Sacred Space
Recommended Reading (Bibliography)
Contacts and Sources (Webworking)
Spells
    Basic Spell Construction and Binding
    Pagan Ritual for Basic Use
    Cleansing a Home
    To Banish Negative Forms from Your Home
    Banishing Spells Part 1
    Banishing Spell Part 2
    Protection Spell
    Herbal Protection Spell
    Luck Spell
    Love Spell
    Weight Loss Magick
    Truth Spell
        Spell of Thanks
    Spells of the Crone
Crafts and Recipes
    Crafts
    Recipes

Periodically, update your BoS or your "for BoS" file will take over your home. Moreover, nothing is more frustrating than to know that you have a copy of a particular invocation and not be able to find it. Plain or fancy, your BoS is useless if it is empty. I find it helpful to have a folder or file to hold material that has yet to find its way into my book(s). A shoebox holds clippings and a special file on my computer holds information, sorted alphabetically. The electronic data is easier for me to update, and I make time to sort it every month.

    This is the technique I use to organize my HDoS information. The key is really to file information on a regular basis and name your files in a logical, informative way. It does you no good to file all your information away, if you cannot find it again when you need it!

B: Document Examples

Following are several documents that I use online. They are examples of what pagans agree to as part of working with others.

/Art of Ritual Class Requirements:

Potential class members must:

1. Have a strong desire to explore Wicca as a spiritual path.

2. Agree that neither the identities of the members, nor any class business, nor any member's personal business, shall be shared outside the class without the express permission of those involved.

3. Have lives that are settled enough to have room, physically and emotionally, for powerful new experiences and personal growth.

4. Have the support of significant others in the decision to become a class member.

5. Complete all class assignments on time.

6. Agree to abide by the attendance requirements.

7. Have schedules that allow participation in most class activities, including rituals and meetings.

8. Agree to use all knowledge shared by the teachers in the spirit of the Wiccan Rede.

9. Make a serious, good-faith effort to get to know all the other class members and build good relationships with them.

10. Support the class on the material plane through contributions of money or work.

11. Begin to create or obtain basic ritual tools (athame, wand, chalice, pentacle, candlesticks, salt, and water bowls) and set up a personal altar at home.

12. Agree to abide by the by-laws of JaguarMoon cybercoven.

13. Understand that participation in the class does not guarantee, nor demand, eventual Dedication into the coven.

/Art of Ritual Class Agreement (to be returned to the instructor):

I wish to join the Art of Ritual class as taught by JaguarMoon cybercoven. As a class member I will:

1. Work hard to learn the Craft for as long as I am associated with the class.

2. Participate in at least 75 percent of the classes, rituals, and other required activities.

3. Meet on a regular basis with my mentor to discuss my growth within the class.

4. Cultivate relationships of cooperation, friendship, and respect with other class members.

5. Follow the Wiccan Rede: "An' it harm none, do as ye will."

6. Honor the Goddess and the God.

7. Work to protect and heal the Earth and Her creatures.

8. Use magick to affect others only with their express consent.

9. Support other class members in their learning, growth, and aspirations.

10. Keep the identity of other class members and friends confidential, except with their permission.

11. Support the work of JaguarMoon with energy, money, or other resources, as appropriate.

12. Give first priority to the needs of my family and livelihood.

/My Responsibilities as Your Teacher:

This document is one I send out to all my students at the beginning of the year so that they can hold me as accountable for my actions as I hold them for theirs. It provides additional support for the nurturing environment I try to create online.

1. To train and nurture your personal abilities and capacities; not to transform you into a reflection of me.

2. To listen to you, with an open mind and loving heart, especially when you criticize me.

3. To be accountable for all decisions that I make.

4. To support you when you disagree with me.

5. To maintain clear lines of authority within the group, reveal hidden agendas, and defuse covert power plays.

6. To keep any commitments that I make to you.

7. To delegate tasks when necessary.

8. To train others to take my place.

9. To honor my personal needs and encourage you to honor yours.

10. To practice courtesy and always act with respect.

11. To create a safe environment for the expression of feelings, while defining boundaries to protect participants.

12. To admit my mistakes and correct them.

/Coven Compact:

A coven compact is a document that simply and clearly states a coven's philosophy and procedures. It may also be called a "charter" or the "by-laws."

/Statement of Purpose:/

JaguarMoon is a cybercoven in the ShadowMoon Tradition.

▸ Our primary focus is to teach the religion of Wicca on the Internet;

▸ Our secondary goal is to worship the Lord and Lady;

▸ Third, we work together as a coven, doing magick, creating community, and sharing knowledge.

Our first and most important Law is: "An' it harm none, do as ye will."

/Our Tradition:/

The ShadowMoon Tradition continues to evolve, but it is based on British Traditional Witchcraft, informed by additional information and beliefs gathered from the American Eclectic Wiccan movement.

/Basic Operating Rules:/

JaguarMoon has several rules that are vital to the coven's strength:

1. Respect. Members must always respect one another's opinions, beliefs, and attitudes. Disagreement is tolerated; questioning is encouraged; but it must be done in a respectful manner. If misunderstandings occur, members are asked to try and "see the other side." Our coven is a practice of discernment, not judgment. Please remember that someone else's view is not right or wrong—only different.

2. Honesty. Honesty is required of all coven members.

3. Privacy. All information shared within the coven is confidential. Anything that we share is not to be shown or repeated to anyone outside the coven, although the creator or creatrix may grant permission to copy the information or to give it to an outsider. All information regarding coven members' names, addresses, telephone numbers, and e-mail addresses is not to be revealed to any nonmember. Violations of privacy are not tolerated. If confidentiality is breached, the offending person is removed from membership.

4. You Are Responsible. Online communication lacks many of the informative nuances of physical meetings. You are the only one who can speak up for yourself; no one else will notice if you are sad, quiet, or having a bad day. If you need something, you must ask. If you do not like how a situation is handled, it is your responsibility to communicate that feeling.

And, if you are asked to do something and you cannot, then you must say so. Personal responsibility is the heart of being a witch and the coven is an ideal place to explore your boundaries and strengths, as you accept and refuse responsibilities.

5. Ritual Attendance. All rituals—Sabbat and Esbat—are to be attended by every member of the coven. We recognize that life and the sometimes unpredictable nature of Internet communication can prevent a member from joining a ritual. However, we expect that every attempt will be made to attend, or that a member who cannot attend will notify the High Priestess or the entire coven in advance.

Any coven member can log the rituals, but cannot share them with either noncoven people, or coven members who did not attend the ritual. If a coven member wishes to receive the text of the ritual outline, it may be available. The High Priest or Priestess will make that decision.

/Coven Laws/

The Laws of this coven are divided into three areas: mundane, physical, and spiritual. The Mundane Laws (also known as the Coven By-Laws) are to be published and available to any person who asks. The Physical and Spiritual Laws are private and shared only among coven members. Exceptions can be made, with the consensus of the Council of Elders.

/Compact Review/

This Compact is an agreement among all coven members. It is also a document that is not set in stone; for, as the world changes, so should we. Each year, near the start of the year (between Litha and Mabon), all coven members shall review and discuss this document. Any changes will be made at that time, and the document signed anew.

Created by Lady Ma'at
High Priestess
Approved by the Council of Elders
June 2000

C: Pagan and Internet Resources

What follows is a collection of my favorite URLs on the Internet and the Web. Use them as a starting place to explore Paganism and Cyberspace. Online information changes quickly and some of these links may no longer be accurate. Try searching for the authors if the link is invalid; it may be that the Web site moved.

http://dir.groups.yahoo.com/dir/Religion_Beliefs/Paganism has the latest **pagan-oriented mailing lists** at Yahoo! Groups. Of those, I highly recommend: A Recipe for Magick; Goddess 2000 Project; All Witches Magic; and Brigid's Well.

http://dmoz.org/Society/Religion_and_Spirituality/Pagan/ for a detailed look at a comprehensive directory of **pagan-related groups**, lists, pages, and services.

http://www.icq.com/icqlist/ReligiousPersuasio/OtherGroups.html for a list of **ICQ magickal groups** and pagans.

http://www.beliefnet.com/ has lots of information on the Divine.

http://www.cauldrons-broomsticks.net/ is a fantastic **pagan e-zine.**

A couple of great **astrology** links can be found at:

http://www.astro.com/ and http://www.astrologyzone.com/.

My favorite **tarot** links are:

http://www.learntarot.com/top.htm

http://www.nccn.net/~tarot/articles.html.

http://babelfish.altavista.com/translate.dyn **translates** anything written—a word, phrase, or paragraph—from one language to another.

For dealing with **Internet security**:

http://www.zonelabs.com/zonealarm.htm. I highly recommend this site, and the program created to block unwanted Internet connections. If you have a cable or DSL connection, you need this program.

http://www.symantec.com/ Norton Anti Virus, or

http://software.mcafee.com/. Get it; your system will thank you. **Virus protection** is a must. Do it...today...now!

Fonts to make your screen look more interesting, or your printouts more elegant can be found at:

http://www.geocities.com/TimesSquare/Alley/1557/fonts1.htm

And, because your **health** is important:

http://www.sfwa.org/ergonomics/eyestrain.htm has great information on ergonomics and eyestrain.

D: Pagan and Internet Glossary

This is a glossary of words commonly associated with both the Internet and Paganism.

/A

Adept: An individual who, through serious study and accomplishments, is considered highly proficient in a particular magickal system.

Air: In most magickal traditions this corresponds with East, the color yellow, the mind, intelligence, and imagination. Other traditions associate it with North.

Akasha: The fifth element, the omnipresent spiritual power that permeates the universe. Also called simply "Spirit" or "ether."

Akashic Records: The complete record of every human experience throughout the entirety of human existence.

Altar: A special, flat surface set aside exclusively for magickal activities or ritual.

Amulet: A magickally charged, protective object that deflects specific, usually negative, energies. It may also act as a "magnet" for specific, usually positive, energies.

American Eclectic Wicca: Sometimes used to refer to the philosophy, rituals, and practices based on the published works of such modern Wiccans as Scott Cunningham and Starhawk. The basic perception is that Wicca is a modern religion created by Gerald Gardner and that the beliefs and practices of Wicca are completely individualistic.

Animism: The belief that natural objects, and nature itself, are alive and conscious; or the belief that an immaterial force animates the universe.

Arcana: Secret or specialized knowledge or detail known only to a few. Also: the two halves of a tarot deck. The Major Arcana consists of twenty-two trumps; the Minor Arcana consists of fifty-six suit cards, and is sometimes called the lesser or lower Arcana.

Aspect: The particular principle or part of the Deity being worked with or acknowledged, usually during ritual or other magickal activities.

Asperger: A ritual tool used to sprinkle consecrated water for purification purposes. It can be a crafted artifact or something as simple as a pinecone, a bunch of fresh herbs, or a twig tied with leaves and needles.

Astral Plane: An invisible parallel world that remains unseen from the physical world.

Astral Projection: The process of separating your astral body from your physical one to travel in the astral plane.

Astrology: The study of and belief in the effects that the movements and placements of planets and other heavenly bodies have on the lives and behavior of human beings. Astrology is frequently used for ritual planning.

Athame: The sacred black-handled knife of the witch. This is traditionally a double-edged knife with a steel blade, often with sacred symbols inscribed on the hilt. They are not used for physical cutting, which is reserved for the Bolline: but for directing power during ritual workings and cutting the sacred Circle. Most magickal traditions associate it with Air, but it can be a symbol for Fire. Generally pronounced "A-tha-may," the term is of obscure origin, has many variant spellings, and an even greater variety of pronunciations.

Attunement: An activity that brings the minds, emotions, and psyches of a group into harmony prior to ritual: chanting, singing, guided meditation, and breathing exercises are common ways to attune.

Aura: The energy field that surrounds the human body, and especially that radiant portion visible to the third eye or psychic vision. The aura can reveal information about a person's health and emotional state.

Auto-Op: The automatic process of making a specific person a high-level user in IRC.

/B

Bandwidth: The amount of data that can be passed along a communications channel in a given period of time.

Banish: To magickally end something or exorcise unwanted energy.

B.C.E.: "Before Common Era," synonymous with B.C. (Before Christ), but without religious bias.

Bell: Often used as a ritual tool; bells can be used to invoke directional energies or clear a space. Associated with the element of Air.

Beltane: A Wiccan festival celebrated on April 30th or May 1st, depending on which tradition your coven follows. Sometimes called May Eve, Roodmas, Walpurgis Night, or Cethsamhain. This fire festival celebrates the symbolic union of the Goddess and God.

Besom: A witch's broom, used for sweeping away energies among other things.

Bi-location: A type of astral projection during which you maintain awareness of your present surroundings.

Bind: To magickally restrain something or someone.

Blood of the Moon: A woman's menses, also called her Moon time.

Bolline: The white-handled knife used in magick and Wiccan ritual for carving or cutting materials that are necessary for ritual or healing. It is also sometimes used for harvesting herbs, and will have a small, silver, sickle-shaped blade.

Book of Shadows (also known as a **BoS**): A witch's book of rituals, spells, dreams, herbal recipes, magickal lore, and so forth; a sort of magickal cookbook. Once hand-copied at Initiation, the BoS is now photocopied, typed, or given via diskette in some covens. No one true Book of Shadows exists; all are relevant to their respective users. The BoS is also called an HDoS, or Hard Disk of Shadows.

Bps (Bits-per-second): A measurement of how fast data ("bits") is moved from one place to another. A 28.8 modem can move 28,800 bits per second.

British Traditional Witchcraft: Traditions tracing descent from a hereditary British source. Includes the Gardnerian traditions and their offshoots, but also several others derived from British sources, such as Sybil Leek's Horsa Coven in the New Forest, Plant Bran, and the Clan of Tubal Cain of Robert Cochrane. The term "British Traditional Wicca" is sometimes used, more commonly in the United States than elsewhere.

Browser: A software program that is used to look at various kinds of Internet resources.

Browsing: Using a browser; also called "surfing the web."

/⊏

CD: Compact disk. A portable data storage medium.

CD-drive: A device to read the data stored on a CD; once known as a CD-ROM drive (ROM stands for Read-Only Memory), but now many of them are able to write to a CD as well as read a CD.

Cable: An information distribution system in which information, picked up by elevated antennas, is delivered by cable to the receivers of subscribers.

Cakes and Ale: The Wiccan "communion" that consists of a natural beverage and cake offered to each participant in a ritual, or eaten by participants at the end of the ritual as a part of the grounding process.

Call: Invoke divine and elemental forces.

Candles: In addition to illuminating the altar, candles are sometimes used to mark each of the four Quarters, and in spell-working. They are often anointed with oil and inscribed with magickal symbols.

Casting the circle: The psychic creation of a sphere of energy around the area where a ritual occurs, to concentrate and focus the power raised, and to keep out unwanted influences or distractions. The space exists between the worlds, outside of space and time.

Cauldron: This is used to make brews, contain a ritual fire, or scrying. Can be four-legged or three-legged. The cauldron represents the womb of the Goddess, as in Cerridwen's cauldron named Aven; or the source of all plenty, as in the Dagda's cauldron.

C.E.: "Common Era," synonymous with A.D., but without religious bias.

Censer: A heatproof container in which incense is burned. Smoke is associated with the element of Air.

Centering: The process of moving one's consciousness to one's spiritual center, leading to a feeling of peace, strength, clarity, and stability.

Ceremonial Magick: A highly codified magickal tradition based on the Kabbala.

Chakra(s): Seven major energy vortexes found in the human body. Each is usually associated with a color.

Chalice: A ritual tool. It represents the female principle of creation and is associated with the West and Water.

Channeling: A practice wherein you allow a discarnate entity to "borrow" your body to communicate.

Chant: This can be a rhyme, sometimes called a rune, intoned rhythmically to raise power. Such rhymes can be simple and repetitive, it makes them easier to remember, but it is not a requirement.

Charge: To infuse an item with magickal energy.

Charge of the Goddess: Originally written in modern form by Doreen Valiente, it is a story of the message from the Goddess to Her children. There are many variations, including Charge of the God, Charge of the Dark Goddess, etc.

Charging: Infusing an object with personal power. Charging is an act of magick.

Charm: An amulet or talisman that has been charged with energy for a specific task.

Circle: Sacred space wherein all magick is worked and ritual energy contained. Also, a gathering of witches or pagans. We meet in circles to worship and work magick. Sometimes the entire group is also known as the circle, grove, or temple.

Cleansing (also called **clearing**): The act of removing negative energies from an object or space.

Collective Unconsciousness: Term used to describe the sentient connection of all living things, past and present; also called the Akashic Records.

Coming-of-Age Ritual: In many traditions, pagan children are seen as spiritual adults at age thirteen for boys, and at the time of a girl's first menses. This ritual celebrates their new maturity and, in some traditions, this is the age when they can hold membership in covens.

Cone of Power: Psychic energy raised and focused during ritual to achieve a definite purpose.

Conscious Mind: The analytical, rational part of our consciousness. This part helps us to balance our checkbooks, theorize, and communicate.

Consecrate: To declare or set apart as sacred; to produce the ritual transformation of an item into something sacred.

Consecration: The act of blessing an object or place by instilling it with positive energy for sacred purposes.

Cookie(s): The most common meaning of "cookie" on the Internet refers to a piece of information sent by a Web server to a Web browser that the browser software is expected to save and to send back to the server whenever the browser makes additional requests from the server. Cookies might contain information such as login or registration information and user preferences, and are usually set to expire after a predetermined amount of time. The information is usually saved in memory until the browser software is closed down, at which time the information may be saved to disk if its expiration time has not been reached.

Cord: May also be called a girdle or cingulum. In many traditions, the color signifies a degree. It may also be used in knot magick, as well as in binding and loosening spells.

Correspondences: A system of symbolic equivalencies used in magick (see Magickal Correspondences).

Coven: A group of witches that work magick or perform rituals together.

Covenstead: The meeting place of witches, traditionally a fixed building or place.

Cowan: A general term indicating anyone who is not a witch, Wiccan, or pagan.

Craft: Short for "witchcraft," or "the Craft of the Wise." Generally associated with the practical aspects of the religion.

Crone: Aspect of the Goddess represented by the Wise Old Woman. Symbolized by the waning Moon, the carrion crow, the cauldron, the color black. Her Sabbats are Mabon and Samhain.

Cross-Quarter Days: Refers to Sabbats falling between the solstices or equinoxes.

Cult: A group that professes to be spiritual in nature, but requires a great deal of energy, money, and obedience from its followers as proof of their devotion. Also frequently dependent on the charisma of a leader.

/D

Days of Power: Days triggered by astrological occurrences, Moon cycles, or special events—your birthday, your menstrual cycle, your dedication or initiation anniversary (see also Sabbat).

Dedication: The ritual in which an individual accepts the Craft as his or her path and vows to study and learn all that is necessary to reach adeptship. It is a conscious preparation to accept something new into your life and stick with it, regardless of the highs and lows that follow. This person may be called a Dedicant, Seeker, or Candidate.

Degrees: Levels of initiation representing spiritual or magickal development, as well as skill, knowledge, and experience (see Initiation).

Deity: A god or goddess.

Deosil: Clockwise, the direction in which the shadow on a sundial moves as the Sun travels across the sky. Deosil is symbolic of life, positive magick, and positive energies.

Divination: The magickal art of discovering the unknown by interpreting random patterns or symbols through the use of devices such as clouds, tarot cards, a bowl of water, or the flame of a candle. Divination contacts the psychic mind by dowsing the conscious mind through ritual and observation, or the manipulation of Wiccan tools.

Divine Power: The unmanifested, pure energy that exists within the Goddess and God. The life force, which is the ultimate source of all living things.

Drawing Down the Moon: A ritual that witches perform during the full Moon to empower themselves and unite their essence with a particular deity, usually the Goddess.

Drawing Down the Sun: A lesser known, and less used, companion ritual to Drawing Down the Moon, in which the essence of the Sun god is drawn into the body of a male witch.

DSL (Digital Subscriber Line): A method for moving data over regular phone lines, which is much faster than a regular dial-up phone connection.

Duality: The opposite of polarity; it separates two opposites such as good and evil and places those characteristics into two completely separate God-forms.

/E

Earth: Revered by pagans and in other Earth-honoring traditions such as Gaia, the Great Mother, who sustains us all as her children. In many magickal traditions, the element corresponds to the North; the colors black, brown, and forest green; foundation, stability, the human body, all solid material things, and prosperity.

Earthing: Sending excess energy into the Earth; done in ritual after power has been raised and sent to its goal (see also Grounding).

Earth Magick or **Power**: The energy that exists within stones, herbs, flames, wind, and other natural objects.

Earth Plane: Metaphor for your normal waking consciousness, or for the world in which we live.

Elder: One who is recognized as an experienced leader, teacher, and counselor.

Elements: Usually Earth, Air, Fire, Water, and Ether in some traditions; they are the building blocks of the universe. Everything that exists contains one or more of these energies.

Elementals: Archetypal spirit beings associated with one of the four elements (see also Quarters).

Emoticon: A series of keyed characters used in text-only environments to indicate an emotion, such as pleasure [:-)] or sadness [:-(|.

Enchantment: A spell.

Equinox: Either of the two times each year, on or about March 21 and September 21, when the Sun crosses the equator and day and night are of equal length.

Esbat: A ritual occurring on a Moon phase (Dark, New, or Full) and usually dedicated to the Goddess in her lunar aspect.

Ether: Another element; the psychic mists out of which inspiration and messages from Spirit can sometimes emerge.

Evocation: To call something out from within.

/F

Familiar: An animal that has a spiritual bond with a witch, many times a family pet. Familiars can also be entities that dwell on the astral plane (see also Totem Animal).

Fascination: A mental effort to control another animal or person's mind; often considered unethical.

Fire: In many traditions, the element corresponds to the South, the color red, energy, will, passion, determination, purpose, ambition, and spirituality.

First Quarter: One half of the Moon appears illuminated by direct sunlight while the illuminated part is increasing. Here the Moon is associated with the Maiden/Mother aspect of the Goddess.

Flame: Originally, "flame" meant to carry forth in a passionate manner in the spirit of honorable debate. Flames most often involved using flowery language and being good at flaming was considered an art. More recently, "flame" has come to refer to any kind of derogatory comment, no matter how witless or crude (see also Flame War).

Flame War: When an online discussion degenerates into a series of personal attacks against the debators, rather than discussion of their positions. A heated exchange.

Folklore: Traditional sayings, cures, wisdom of a particular locale which is separate from their mythology.

Folk Magick: The practice of projecting personal power, as well as the energies within natural objects such as herbs and crystals, to bring about needed changes.

FTP (File Transfer Protocol): A very common method of moving files between two Internet sites. FTP is a special way to log in to another Internet site for the purposes of retrieving and/or sending files. There are many Internet sites that have established publicly accessible repositories of material that can be obtained using FTP, by logging in using the account name "anonymous," thus these sites are called "anonymous FTP servers."

Full Moon: The fully visible Moon, associated with the Mother aspect of the Goddess.

/G

Gaea/Gaia: Great Mother Earth, who sustains us all as her children with her love and abundance.

Gardnerian: Gardnerian Witchcraft is the tradition taught by Gerald Gardner and his initiates, primarily as it was passed on from his original coven.

Gigabyte (GB): 1000 or 1024 MB, depending on who is measuring.

God: Masculine aspect of Divinity (see also Lord).

Goddess: Feminine aspect of Divinity (see also Lady).

Great Rite: Symbolic sexual union of the Goddess and God that is generally enacted at Beltane in many traditions. It symbolizes the primal act of creation from which all life comes. It may also be used to bind a person to the land, in which case it is frequently referred to as the sacred marriage. Please note the word "symbolic." The Great Rite is rarely (if ever) an actual sexual union in modern magickal groups.

Green Man: Another name for the God, referring to his aspect as Lord of the Woodlands.

Grimoire: A magickal workbook containing ritual information, formulae, magickal properties of natural objects and preparation of ritual equipment. Often used interchangeably with Book of Shadows.

Grounding: The process of connecting oneself to the Earth. It is absolutely necessary to ground and center before one does any work involving energy or power. If you do not, you are likely to have a headache at the end of a ritual. You may also faint or become quite ill.

Guardians: The creatures or angels that protect the four quadrants or elements of the Circle.

/H

Handfasting: A pagan wedding.

Healing: The goal of a great deal of magick. Some alternative forms of healing include chakra or energy work, visualization, herbcraft, and spirit journeys, among many others. Many witches are professionals in the fields of healing and medicine.

Heathen: A non-Christian, from "one who dwells on the heath."

Herb: A plant or plant part valued for its medicinal, savory, or aromatic qualities.

Herbalism: Art of using herbs both magickally and medicinally to facilitate human needs.

High Priest: The primary male leader within a coven. Usually a 3rd-degree initiate who either helped found the coven, was chosen by the High Priestess, or was elected by the members.

High Priestess: The primary female leader within a coven. Usually a 3rd-degree initiate who either founded the coven, or was elected by the members.

Higher Self: That part of us that connects our corporeal minds to the collective unconscious and with the divine knowledge of the universe.

Hiving Off: The process in which a coven splits off part of itself to form a new, separate entity, generally within the same tradition. Although sometimes this is done to keep the coven a manageable size, it may also occur when there are irreconcilable philosophical or political differences.

Horned God: One of the most prevalent god-images in Paganism.

/I

I Ching: A Chinese system of divination in which yarrow stalks or coins are cast to create hexagrams, which are then interpreted from a standard I Ching book, such as *The I Ching or Book of Changes* by Richard Wilhelm.

Illuminati: Whenever conspiracy theory is spouted, the mysterious "Illuminati" are most often named as being responsible. Ironically, while people can name those ostensibly belonging to other conspiracy groups, the "Illuminati" is always left hanging as some secret, shadowy entity that no one can quite describe. Interestingly, too, no one can quite identify what specific acts can be attributed to them, and no one in 225 years seems to have left the organization to reveal its secrets.

Imbolc: A Wiccan festival celebrated on February 2, also known as Candlemas, Lupercalia, Feast of Troches, Oimelc, Brigit's Day, and other names. Imbolc celebrates the first stirrings of spring and the recovery of the Goddess from giving birth to the Sun God at Yule.

Immanence: The belief that Deity exists within all things, including people, and cannot be separated from them.

Incense: Ritual burning of herbs, oils, or other aromatic items to scent the air during magick and ritual, and to help the witch attune his or her energy. A symbol of Fire or Air, they may be in the form of sticks, cones, resins, or dried herbs. The incense chosen depends on the type of magick. If a witch has difficulty breathing when incense is burned (such as in cases of asthma or other breathing diseases), the use of essential oils is a perfectly acceptable substitute. I recommend placing three drops of oil in three tablespoons of water in a small dish over a tea candle.

Initiate: Someone who has been through a ritual of Initiation and taken vows. These vows mostly are to protect the Craft and other practitioners, and dedicate oneself to the gods and goddesses. They vary a bit from group to group. Solitary practitioners may be self-initiated witches.

Initiation: A ritual during which an individual is introduced or admitted into a coven; not to be confused with Dedication. Coven initiations are generally associated with degrees. The question of whether a self-initiation is valid is still under discussion in the pagan community. In my opinion, "The gods and goddesses initiate. We just officiate." But only covens use degrees.

Internet: The vast collection of interconnected networks that use the TCP/IP protocols, and that evolved from the ARPANET of the late 1960s and early 1970s. The Internet connects well over 60,000 independent networks into a vast global network.

Invocation: An appeal or petition to a higher power or powers, such as the Goddess and God. Invocations take the form of a spoken prayer at some point in a ritual. Invocation is actually a method of establishing conscious ties with those aspects of the Goddess and God that dwell within us.

Invoke: To bring something in from without.

ISP (Internet Service Provider): An institution that provides access to the Internet in some form, usually for money.

/K

Kabbala: Mystical teaching from the Jewish-Gnostic tradition. Ceremonial Magick and the Gardnerian traditions are based in these teachings. Also spelled "Qabala."

Karma: The belief that one's thoughts and deeds can either be counted against one or added to one's spiritual path across several lifetimes.

Kilobyte (KB): Literally means one thousand bytes, but in technical terms, it's 1024 bytes.

/L

Labrys: A double-headed ax that symbolizes the goddess in Her lunar aspect. Although its roots are found in ancient Crete, many Dianic witches use this symbol in their practice.

Lady: The Goddess. We refer to Her by any of the names of Her aspects. She is a triple Goddess: Maiden, Mother, and Crone. We identify ourselves with different aspects of Her at different points in our lives. We also appeal to different aspects of Her for different requests. She is all-encompassing, containing especially all creative and procreative properties: nurturing and healing, birth and rebirth, and growth and plenty. All women embody Her nature and are part of Her. All living things are Her children. She is the Earth, the Moon, the stars, and the rain. Together, the Lord and the Lady bring forth all life that is on the Earth.

Last Quarter: One half of the Moon appears illuminated by direct sunlight while the illuminated part is waning. Here, the Moon is seen as the Crone aspect of the Goddess.

Law of Return: Whatever energy is sent out returns to the sender multiplied. Some traditions say it is multiplied by three and, therefore, call this principle the "Threefold Law."

Laws of Witchcraft: A list of rules for witches that focus on individual conduct and coven operations. They are sometimes called the Ordains. Several versions exist. Their origins are unclear: they may be from the Burning Times, or more recent, or a pastiche of ancient and modern. Generally, each coven has a set of by-laws, sometimes divided into spiritual, physical or mundane, and magickal categories.

Libation: Ritually given portion of food or drink to a Deity, nature spirit, or ghost.

Lord: The God. We refer to Him by the names of His aspects. He is the Horned God of the Hunt, or the Lord of Death and Resurrection. We often refer to Him as the Laughing Lord. He is the Lady's Consort, who dies and is reborn each year. He is sensuality, strength, music, and lust. All men embody His nature and are part of Him. Together, the Lord and the Lady bring forth all life that is on the Earth.

Lughnasadh (pronounced LOO-nah-sah): A Wiccan festival celebrated on August 1, also known as August Eve, Lammas, or the Feast of Bread. Lughnasadh marks the first harvest, when the fruits of the Earth are cut and stored for the dark winter months, and the God weakens as the days grow shorter.

Lunar Cycle: An approximate twenty-nine-day cycle during which the visible phase of the Moon waxes from dark to full and wanes to dark again. Much magick is geared to the energies present at certain phases of the lunar cycle.

Lurking: The very common practice of reading an online or e-mail discussion without taking part in it.

/M

Mabon: On or around September 21, the Autumn Equinox, Wiccans celebrate the second harvest. Nature is preparing for winter.

Macrocosm: The world around us.

Magick: Using knowledge of the Craft and mental focus to direct energy and manifest a change in physical reality.

Magick Circle: A protective sphere delineated at the beginning of a ritual in which magick is performed. The sphere extends both above and below the surface of the ground.

Magickal Correspondences: Items, objects, days, colors, Moon phases, oils, angels, and herbs used in a ritual or magickal working that support the intent or purpose.

Magickal System: The basic set of guidelines relating to the worship of specific gods and goddesses, or cultural traditions. Also called a tradition.

Maiden: Youngest aspect of the triple goddess. Represented by the waxing Moon, colors white and blue. Her Sabbats are Imbolc and Ostara.

Male Mysteries: Pagan study that attempts to reclaim the power and mystery of the old gods for contemporary pagan males.

MBps (Megabytes-per-second): A reference to how fast information can be transmitted to and from computers and servers.

Meditation: Reflection, contemplation, turning inward toward the self, or outward toward a deity or nature. A quiet time in which the practitioner may dwell on particular thoughts or symbols, or allow them to come unbidden.

Megabyte (MB): Literally means one million bytes, but is technically 1024 kilobytes.

Microcosm: The world within us.

Midsummer: The Summer Solstice, also known as Litha. Usually falls near June 21. One of the Wiccan festivals and an excellent night for Magick. Midsummer marks the point of the year when the Sun is symbolically at the height of its powers, and so too is the God. This is the longest day of the year.

Monotheism: Belief in one supreme deity who has no other forms or aspects.

Mother: The aspect of the Goddess that represents motherhood, mid-life, and fertility. She is represented by the full Moon, the egg, the colors red and green. Her Sabbats are Litha and Lughnasadh.

Myth: Body of lore about a recurring story, handed down through time, that is indigenous to any land or people that illustrates the "why" behind the change in seasons, phases of the Moon, or other natural phenomena.

/N

Neopagan: Literally, a new pagan. General term for contemporary follow-ers of Wicca and other magickal, shamanistic, and polytheistic Earth-hon-oring religions.

Netiquette: The etiquette of online communication and behavior.

New Age: The current mixing of metaphysical practices with a structured spiritual practice.

New Moon: The Moon when it is not illuminated by direct sunlight. The Moon as Maiden.

/O

Occult: Literal meaning is "hidden" and is broadly applied to a wide range of metaphysical topics that lie outside the accepted realm of mainstream theologies.

Occultist: One who practices and or studies a variety of esoteric subjects.

Old Ones: Refers to all aspects of the Goddess and God.

Old Religion: Another name for Paganism, Wicca, or other Earth-honoring practices.

Operating System: Software designed to control the hardware of a specific data-processing system in order to allow users and application programs to make use of it.

Ostara: Occurring at the Spring Equinox, around March 21, Ostara marks the beginning of true, astronomical spring, when snow and ice make way for green. As such, it is a fire and fertility festival, celebrating the return of the Sun—the God—and the fertility of the Earth—the Goddess.

/P

Pagan: From the Latin *paganus*, meaning "country dweller." One who practices a religion outside of the mainstream (Judaism, Christianity, Muslim, Buddhism). All witches are pagan; but not all pagans are witches.

Paganing: When a baby is presented in circle to the Goddess and God, and given a Craft name that he or she will keep until about the age thirteen. Also called Wiccaning.

Pantheon: A collection or group of gods and goddesses in a particular religious or mythical structure.

Pantheism: Belief in many deities. Paganism is pantheistic.

Passing Over Ritual: Ritual observed when a loved one has died.

Past Life Regression: Act of using meditation or guided meditation to pass through the veil of linear time and perceive experiences encountered in a previous existence.

Path Working: Using astral projection, bi-location, or dreamtime to accomplish a specific goal. The Native Americans called it vision questing.

Pentacle: A circle surrounding a five-pointed, upright star (pentagram). Often worn as a symbol of a witch's beliefs. The circle represents unity or the world.

Pentagram: The basic interlaced five-pointed star, visualized with one point up. It represents the five elements: Earth, Air, Fire, Water, and Spirit. It is a symbol of power and protection.

Personal Power: The energy that sustains our bodies. It originates within the Goddess and God. We first absorb it from our biological mother in the womb, and later from food, water, the Moon and Sun, and other natural objects.

Planetary Hours: A system of hourly divisions associated with planetary energies.

Polarity: The concept of equal, yet opposite, energies. The Eastern yin/yang is a perfect example. Yin is cold; yang is hot. Other examples: goddess/god, night/day, Moon/Sun, birth/death, dark/light, psychic mind/conscious mind. Universal balance.

Polytheism: Belief in the existence of many deities.

Priest: A male dedicated to the service of his chosen deity, his coven, and humankind. All male Wiccans are priests after they have been initiated (self-initiation or by a coven). In a coven setting, a High Priest may act as an administrator for the coven and a co-leader in rituals.

Priestess: A female dedicated to the service of her chosen deity, her coven, and humankind. All female Wiccans are priestesses after they have been initiated (self-initiation or by a coven). In a coven setting, a High Priestess may act as leader of the coven and in rituals.

Processor: A program that translates another program into a form acceptable by the computer being used.

Projective Hand: The dominant hand through which personal power is sent from the body. Normally the hand used for manual activities such as writing or dialing the telephone, for instance.

Psychic Mind: The subconscious or unconscious mind, in which we receive psychic impressions. It is at work when we sleep, dream, and meditate. It is our direct link with the Divine, and with the unseen world.

Psychic trance: When a person enters the altered state of consciousness in which he or she is no longer using the normal *beta* waves associated with ordinary consciousness, but instead the *theta* and *delta* waves associated with sleep. A witch enters this state to do acts of magick.

Psychism: The act of being consciously psychic, in which the psychic mind and conscious mind are linked and working in harmony. Also known as psychic awareness.

/Q

Qabala: See Kabbala

Quarters: The cardinal directions, corresponding to the elements and protected by the guardians, or the Sabbats that fall on the equinoxes or solstices.

/R

RAM (Random Access Memory): A memory device in which information can be accessed in any order. The memory your computer needs to run software programs.

Receptive Hand: The hand through which energy is received into the body. The left hand in right-handed persons, or the reverse for left-handed persons.

Rede: "An' it harm none, do what ye will." The most basic tenet of witchcraft.

Reincarnation: The process of repeated incarnations in human form to allow evolution of the sexless, ageless soul. The belief that souls do not end at death, but wait for a time and then are reborn to live and learn on this Earth again. This is also one of the central tenets of the Craft.

Ritual: A specific form of movement, a manipulation of objects, or inner processes designed to produce metaphysical effects. The goal of ritual is union with the Divine. Magickal works are performed to produce a specific state of consciousness that allows the witch to focus and move energy.

Ritual Consciousness: A specific, alternate state of awareness necessary to the successful practice of magick. This state is achieved through the use of visualization and meditation. The conscious mind becomes attuned with the psychic mind; in this state, the witch senses energies, gives them purpose, and releases them toward a specific goal. It is a heightening of senses, an expanded awareness of the nonphysical world, a linking with nature and Deity.

Ritual Robe: A ritual robe is a standard piece in the pagan wardrobe. Worn when at ritual or circles, or worn while practicing as a solitary, the robe is a spiritual garment.

Ritual Tools: General name for magickal tools used by a witch or magician. They vary by tradition and each usually represents one of the elements.

Runes: A set of symbols used in divination and magickal work; derived from an ancient Anglo-Saxon alphabet.

/S

Sabbat: One of the eight yearly holidays that Wiccans celebrate. A time for feasting, partying, and merrymaking. The word Sabbat is given various derivations, but I prefer the argument that it comes from the French *s'e-battre*, which means, "to frolic" (see also Wheel of the Year).

Samhain (pronounced SOW-en): A Wiccan festival celebrated on October 31, also known as November Eve, Hallowmas, Halloween, Feast of Souls, Feast of the Dead, Feast of Apples. Samhain marks the symbolic death of the Sun god and His passing into the "land of the young," where he awaits rebirth at Yule. This Celtic word is pronounced by Wiccans as: SOW-wen, SEW-wen, SAHM-hain, SAHM-ain, SAV-een, and other ways. The first seems to be the one preferred among most Wiccans. This festival is considered to be the Witches' New Year.

Scourge: Small device made from leather or hemp that resembles a multi-stranded whip. Specifically associated with the Gardnerian tradition, the scourge is not designed to cause pain in any way, and is completely symbolic.

Scrying: A method of divination. To gaze at or into an object such as a crystal ball, a pool of water, mirrors, or a candle flame, while stilling the conscious mind in order to contact the psychic mind. Scrying allows the practitioner to become aware of events prior to their actual occurrence, as well as to perceive past or present events through other than the five senses.

Search Engine: A software program that searches a database and gathers and reports information that contains or is related to specified terms. Also, a Web site that functions primarily as a searching device that gathers and reports information available on the Internet or a portion of the Internet.

Shaman: A man or woman who has obtained knowledge of the subtler dimensions of the Earth, usually through periods of alternate states of consciousness. Various types of ritual allow the shaman to pierce the veil of the physical world and to experience the realm of energies. This knowledge lends the shaman the power to change his or her world through magick.

Shamanism: The practice of shamans, usually ritualistic or magickal in nature, sometimes religious.

Shrine: A sacred place that holds a collection of objects that represent a deity.

Sky-Clad: Naked.

Solitary: Pagan who does ritual alone.

Solstice: The time of year when the Sun crosses the equator during its annual trip from north to south as it travels along its yearly orbit. After the Summer Solstice, the days grow shorter; they grow longer after the Winter Solstice.

Spam: Inappropriate use of a mailing list, USENET, or other networked communications facility as a broadcast medium (which they are not) by sending the same message to a large number of people who didn't ask for it. The term comes from a famous Monty Python skit that featured the word "spam" repeated over and over. The term may also have come from someone's low opinion of the food product with the same name, which is generally perceived as a generic content-free waste of resources.

Spamming: The act of sending spam to an e-mail list.

Spell: A magickal ritual, often accompanied by spoken words to direct energy toward a goal. It should be ethical in purpose, clear, concise, focused, and emotional. A spell can be written, spoken, or drawn. The best spells are the ones you write yourself; but no spell will work unless true need is present. Also, magick always takes the path of least resistance; a spell can only open a path that you must take to reach your goal.

Spiral: This symbol signifies an inward journey. It represents the emergence into consciousness of what was previously hidden. It also suggests the round of seasons, where life unfolds and fades, unfolds again in a repeating cycle—it means life. For many witches, the spiral is a symbol of the Goddess.

Spiral, Double: The double spiral is a symbol of the witch's descent into the Underworld, and return from death into life.

Spirit: The overall energy that runs the universe in a harmonious way. The fifth element.

Staff: Ritual tool corresponding to the wand, although seldom used.

Subconscious Mind: Part of the mind that functions below the levels that we access in the course of a normal working day. This area stores symbolic knowledge, dreams, and the minutest details of every experience ever had by a person.

Summerland: The pagan Land of the Dead.

Sympathetic Magick: Concept of like attracts like. The most common way that spells are worked.

/ T

Talisman: An object charged with personal power to attract a specific force or energy, or to protect its bearer.

Tarot (Cards): Set of cards that features pictures and symbols; used to connect the diviner with the collective unconscious. The cards frequently reveal previously unknown information or can help predict future events.

Threefold Law: Karmic principle that energy released is returned three times over.

Totem Animal: A spiritual guide in the form of an animal that embodies qualities that the seeker wants to attain or enhance.

Tradition: Branch of Paganism followed by any individual pagan or coven. Examples include: Gardnerian, American Eclectic Wicca, and British Traditional Wicca.

Trance: An altered state of consciousness.

Triple Goddess: The Goddess in all of her three aspects: Maiden, Mother, Crone.

/ V

Vision Quest: In modern times, this frequently refers to the process of using astral projection, bi-location, or dreamtime to accomplish a specific goal. Also called path-working. Its origins are shamanistic, and are based on a Native American practice. A member of the tribe would go into the wilderness to seek spiritual direction.

Visualization: The process of forming mental images for the purpose of shifting energy. Magickal visualization consists of forming images of goals during ritual. It is also used to direct personal power and natural energies for various purposes during magick, including casting the circle.

/ W

Wand: Ritual tool representing Fire.

Waning Crescent: The Moon is partly, but less than one-half, illuminated by direct sunlight while the illuminated part is waning. In this phase, the Moon represents the Crone.

Waning Gibbous: The Moon is less than fully, but more than one-half, illuminated by direct sunlight while the illuminated part is waning. Three days after the Full Moon. The Moon here is Mother.

Waning Moon: A phase of the Moon in which the face of the Moon is getting smaller. The time between a Full Moon and a New Moon.

Water: In most magickal traditions, this corresponds with West, the color blue, the psychic mind, intuition, and emotion.

Waxing Crescent: The visible Moon is partly, but less than one-half, illuminated by direct sunlight while the illuminated part is increasing. In this phase, the Moon represents the Maiden.

Waxing Gibbous: The Moon is more than one-half, but not fully, illuminated by direct sunlight while the illuminated part is increasing. Three days before the Full Moon. In this phase, the Moon represents the Mother.

Waxing Moon: The phase of the Moon in which the face of the Moon is getting larger. The time between a New Moon and a Full Moon.

Web Weaving: Networking with other magickal people to gather information of mutual interest.

Wheel of the Year: The full cycle of the eight Sabbats in the Wiccan calendar. They occur at the Equinoxes and Solstices (the Quarters) and on the days marking the midpoints between them (the Cross-Quarters.) A short list of the names used by my tradition:

Name(s)	Dates
Samhain (pronounced Sowen)	October 31
Yule	Winter Solstice
Candlemas or Imbolc	February 2
Eostar	Spring Equinox
Beltane	May 1
Litha	Summer Solstice
Lughnasad	August 2
Mabon	Autumn Equinox

Wicca: A modern pagan religion that expresses reverence for nature and honors the Earth. Also included are: reverence for the Goddess and God, acceptance of reincarnation and magick, ritual observance of astronomical and agricultural phenomena, and the use of magickal circles for ritual purposes.

Wiccan Creed: This is the moral code of all who practice Wicca and most who practice witchcraft; also known as the Rede: "Eight words the Wiccan Rede fulfill: An' it harm none, do what ye will."

Wicce: Synonymous with Wicca. In some circles, Wicce is used for women and Wicca is used for men. Plural of "witch."

Widdershins: Counterclockwise motion, usually used for inward-looking magickal purposes, or for dispersing negative energies or conditions, such as a disease state present in the physical body.

Witch: A practitioner of the Craft of the Wise; this term applies equally to males or females. Used by some Wiccans to describe themselves.

Witchcraft: The Craft of the witch, including magick, especially magick utilizing personal power in conjunction with the energies in stones, herbs, colors, and other natural objects.

Witches' Pyramid: A creed and a structure of learning that witches follow: "To Know, To Dare, To Will, and To Be Silent."

/Y

Yule: A Wiccan festival celebrated on or about December 21, marking the rebirth of the Sun God from the Earth Goddess. It is a time of joy and celebration during the miseries of winter. Yule occurs on the Winter Solstice.

Endnotes

CHAPTER 1

1. Jennifer J. Cobb, *CyberGrace: the Search for God in the Digital World* (New York: Crown Publishers, 1998), p. 9.
2. Erik Davis, *Techgnosis: Myth, Magic, and Mysticism in the Age of Information* (New York: Crown Publishers, 1998), p. 187.
3. Michael Compton, personal communication.
4. Arthur C. Clarke, *Profiles of the Future: An Inquiry into the Limits of the Possible* (Popular Library, 1977).
5. Aleister Crowley, *The Equinox,* vol. I, no. II (London: Simpkin, Marshall, Hamilton, Kent & Co. Ltd., 1909).

CHAPTER 4

1. Starhawk, *Spiral Dance* (New York: Harper & Row, 1979), p. 35.

CHAPTER 6

1. Dion Fortune, *Applied Magic* (York Beach, ME: Samuel Weiser, 2000).
2. Fortune, *Applied Magic.*
3. Robert A. Heinlein, *The Notebooks of Lazarus Long,* D. F. Vassallo, illustrator (New York: Putnam's Son, 1978).
4. Heinlein, *The Notebooks of Lazarus Long.*

CHAPTER 8

1. This meditation is used with the permission of Briana Ashling.

Bibliography

Adler, Margot. *Drawing Down the Moon: Witches, Druids, Goddess-Worshippers, and Other Pagans in America Today.* Boston: Beacon Press, 1986.

Baldwin, Christina. *Calling the Circle: The First and Future Culture.* New York: Bantam Books, 1994.

Carnes, Robin Deen and Sally Craig. *Sacred Circles: A Guide to Creating Your Own Women's Spirituality Group.* San Francisco: HarperSanFrancisco, 1998.

Cobb, Jennifer. *Cybergrace: The Search for God in the Digital World.* New York: Crown Publishers, 1998.

Crowley, Vivianne. *Wicca: The Old Religion in the New Millennium,* London: Thorsons, 1996.

Cunningham, Scott. *Living Wicca: A Further Guide for the Solitary Practitioner.* St. Paul, MN: Llewellyn Publications, 1995.

———. *Wicca: A Guide for the Solitary Practitioner.* St. Paul, MN: Llewellyn Publications, 1994.

Davis, Erik. *Techgnosis: Myth, Magic and Mysticism in the Age of Information.* New York: Crown Publishers, 1998.

Harrow, Judy. *Wicca Covens: How to Start and Organize Your Own.* Secaucus, NJ: Citadel Press, 1999.

K, Amber. *Covencraft: Witchcraft for Three or More.* St. Paul, MN: Llewellyn Publications, 1998.

Mariechild, Diane. *Mother Wit: A Guide to Healing & Psychic Development.* Freedom, CA: Crossing Press, 1981.

———. *The Inner Dance: A Guide to Spiritual and Psychological Unfolding.* Freedom, CA: Crossing Press, 1987.

McCoy, Edain. *Inside a Witches' Coven,* St. Paul, MN: Llewellyn Publications, 1997.

Morwyn. *Secrets of a Witch's Coven.* Atglen, PA: Whitford Press, 1988.

Pollack, Rachel. *The Power of Ritual.* New York: Dell Publishing, 2000.

Ravenwolf, Silver. *To Ride a Silver Broomstick: New Generation Witchcraft.* St. Paul, MN: Llewellyn Publications, 1994.

———. *To Stir a Magick Cauldron: A Witch's Guide to Casting and Conjuring.* St. Paul, MN: Llewellyn Publications, 1996.

Roth, Gabrielle. *Maps to Ecstasy: Teachings of an Urban Shaman.* San Rafael, CA: New World Library, 1989.

———. *Sweat Your Prayers: Movement as Spiritual Practice.* New York: Putnam Publishing Group, 1997.

Starhawk. *Dreaming the Dark: Magic, Sex and Politics.* Boston: Beacon Press, 1988.

———. *The Spiral Dance: A Rebirth of the Ancient Religion of the Great Goddess.* San Francisco: HarperSanFrancisco, 1979.

———. *Truth or Dare: Encounters with Power, Authority and Mystery.* San Francisco: HarperSanFrancisco, 1987.

Stein, Diane. *Casting the Circle: A Women's Book of Ritual.* Freedom, CA: Crossing Press, 1990.

Simms, Maria Kay. *Witches Circle.* St. Paul, MN: Llewellyn Publications, 1998.

Telesco, Patricia. *The Urban Pagan: Magickal Living in a 9-to-5 World.* St. Paul, MN: Llewellyn Publications, 1993.

Wolfe, Amber. *In the Shadow of the Shaman: Connecting with Self, Nature, and Spirit.* St. Paul, MN: Llewellyn Publications, 1988.

Index